LIGHT,
LASERS
and OPTICS

Notice

Neither the author, the publisher, nor any company or individual mentioned herein makes any warranty, express or implied, or assumes any legal liability or responsibility for any use of the information, methods, and products described in this book or for the results of such use or for the possible infringement of privately owned rights by anyone making such use. The appearance of patented ideas in this book does not release the rights to others. This book cannot give all possible safety precautions for the experimental methods described completely or incompletely herein, and the reader is advised of his/her own responsibility for safety.

LIGHT, LASERS and OPTICS

JOHN H. MAULDIN

TAB BOOKS Inc.
Blue Ridge Summit, PA

FIRST EDITION
FIRST PRINTING

Library of Congress Cataloging in Publication Data

Mauldin, John H.
Light, lasers, and optics / by John H. Mauldin.
  p.   cm.
Bibliography: p.
Includes index.
ISBN 0-8306-9038-7   ISBN 0-8306-9338-6 (pbk.)
1. Light. 2. Optics. 3. Lasers. I. Title.

QC358.M38 1988   88-17068
535—dc19   CIP

TAB BOOKS Inc. offers software for
sale. For information and a catalog,
please contact TAB Software Department,
Blue Ridge Summit, PA 17294-0850.

Questions regarding the content of this book
should be addressed to:

Reader Inquiry Branch
TAB BOOKS Inc.
Blue Ridge Summit, PA 17294-0214

Cover photograph courtesy of Coherent Inc.

Contents

Acknowledgments

Appreciation is expressed to Eugene Hecht and Alfred Zajac, authors of *Optics*, a text published by Addison-Wesley Publishing Company, for showing that advanced optics—even the mathematical description—can be presented in exciting ways. This was an inspiration for attempting the translation of advanced treatments to elementary levels.

Appreciation is expressed to *Scientific Ameri-can* magazine for many decades of outstanding and inspiring visual and textual coverage of topics in modern optics, including the whole issue of September 1968 on light.

Appreciation is expressed to Brint Rutherford, editor at TAB BOOKS, for encouraging this exciting writing project.

Introduction

This book is intended for the general reader interested in the science and technology of light and optics. The level is introductory, and the reader is not assumed to have studied physics previously. The potential audience is broad, ranging from artists interested in the physical foundations of the visual arts to those interested in biology, to physics students wanting a survey of optics, to astronomy enthusiasts, and to those interested in optical technology and media. The basis is given for natural phenomena that are observable nearly every day—for example, various colored spots and rings that occur in conjunction with the sun—as well as for many optical devices and technological advances.

Most of the book is suitable for popular courses in the physics of light and color or in the physics of the visual arts. Designers and artists need to understand the rudiments of how light travels and much about vision, color, and the interaction of light with surfaces. Background is provided for study of drawing, painting, photography, printing, and electronic media. Biologists and psychologists at all levels need

to know about light and optics to study visual perception. Technologists will want to learn of the many uses of the laser and how it is revolutionizing science and technology far beyond optics. Practical philosophers interested in visual information will find the foundations of the subject here.

A tendency when you think about optics might be to think of mirrors, lenses, and other elementary and long-used technology. Popular conceptions of modern optics might center around lasers and holograms. To understand modern topics, you must first understand basic classical optics. The laser is a recent invention. Sophisticated modern physics is involved—although not so sophisticated that it cannot be explained with very simple diagrams. Optical scientists have added other new ideas, like squeezing and doubling light, and filtering spatial frequencies. Biologists are unraveling the complex chemistry of sight and the neural circuitry that analyzes visual patterns. Plants and animals use light in other fascinating ways, too. Philosophers of science are still puzzled about how a photon of light received in one place seems to know all about a photon received elsewhere far away; this is a subject at the outer limits of this book.

Optics is defined here as a science that studies the properties of light as it travels through space and interacts with matter. Light is a form of energy, and *physics* is often said to be the science of energy in all its forms. The production of light involves a conversion of energy from other forms and therefore brings in other branches of physics. The absorption and consequent elimination of light is simply a conversion to another form of energy, too. For those interested in physical science as science, the fundamental methods of science and the laws of physics pertaining to optics are provided. Optics is a very visual science and its experimental basis is made readily apparent. However, there are phenomena that were never suspected to exist until predicted by general theories in optics.

The history of the science of optics is sketched briefly. The main treatment is based on modern approaches, translated to nonmathematical terms. The early student in physics can gain a feel for the content and direction of the field, including topics that would be in a graduate course. Related branches of physics such as electromagnetism, atomic theory, and quantized physics are inseparable from optics and are explained to the extent that an understanding of optics and its applications depend upon them. For the very practical reader, simple observations and home experiments are suggested to provide immediate access to some quite unexpected behaviors of light in nature. Many curious phenomena can be seen using common materials without laboratory equipment.

As a technology, *optics* might be defined as the manipulation of light to achieve various goals. Major applications center around information transmission and processing, and around the generation and detection of light. Many common optical devices and some sophisticated modern instruments are described. Exciting uses of optics, often featuring laser light, are given for many fields, from precision studies of materials to astronomical measurement. A special effort is made to show the breadth of current research and development, although there is space to mention only a fraction of recent contributions. Some of the expected applications are discussed where progress has been thwarted for many years—optical computers, for example. The understanding and use of optics can be expected to expand rapidly in the near future as it has in the past.

Historically, three different general theories of light have been developed, resulting in models of light using rays, particles (corpuscles or photons), and waves. Each has different implications for optics, and each can give useful answers within the limitations of the model. The *ray theory*, which treats light simply as something that travels in straight lines, is now known as *geometric optics* and is at least somewhat familiar to anyone who has worked with lenses. Geometric optics has been subsumed into both the photon and wave models. Optics gained firm theoretical foundations when rudimentary theories of electricity and magnetism were combined in the late nineteenth century to predict the very existence of, and many properties of, light. This book begins with the wave approach and presents as a special case geometric optics involving lenses, mirrors, and so forth. The quantum (photon) theory of light is provided later as a foundation for many modern applications.

Light and optics as discussed here will be limited to a small part of the spectrum of electromagnetic waves. Although some use the term more broadly, optics is commonly understood to pertain to light from the infrared through the visible and into the ultraviolet range. Much of this range is readily generated by lasers and readily controlled by ordinary sorts of lenses and mirrors. Outside this range, radically different methods are needed to manipulate electromagnetic waves, although the basic principles remain the same. Instruments such as the electron microscope are not based on visible light optics and so are omitted even though all microscopes have the ultimate goal of providing visible pictures.

This book should be fully accessible to anyone who has studied some science and math at the high-school level, but some high school physics and/or chemistry might be helpful. The occasional use of mathematics is primarily to provide mathematical symbols as convenient shorthand. Simple mathematical relations are introduced as an option to show how physical laws are expressed and used mathematically. The equivalent is given in words so that continuity is not lost. Occasionally, more advanced mathematical descriptions are given in separate sections in some chapters. All mathematics is accessible below the level of calculus despite the heavy dependence of modern optics on advanced calculus. The ambitious reader is provided with an array of useful physical laws in mathematical form so that estimates or exact calculations of phenomena can be carried out.

Explanation of the basic unit system of science is given, including specialized optical units. Some numerical estimates are given as examples. There is frequent use of numerical angles in degrees (°), minutes, and seconds of arc, and occasional use of radians. Some numbers in optics are very large or small, and these numbers are given in powers of ten—for example, one-trillionth is 10^{-12}. Unusual prefixes such as nano- are defined when first used with units such as meters (nanometer). The reader is assumed to have at least a slight acquaintance with the names of chemical elements.

New and important terms in optics are given in italics where they are defined. The large list of references provides articles and books for further reading. Suggested home observations and experiments are labeled. Possible experimental outcomes are described. Participation is the best form of learning, but experiments can be skipped without loss of continuity.

This book provides information for experimentation beyond the level of experiments given. Because the eye is extremely sensitive and certain light sources such as lasers, the Sun, and mercury lamps are very intense, extreme caution is required. There are many ways accidental overexposure can occur. Ultraviolet and infrared light, being invisible, can cause permanent injury before pain is felt. The eye should not be thought of as a test instrument until all possibility of injury has been eliminated. Sufficient discussion is given that the careful reader should be aware of the dangers in optics, but there is no replacement for experience and guidance when experimenting with light.

1

Light

Optics concerns light. To begin, you will need a simple working description of light. Several will be presented, but details on each will require several chapters to develop. This chapter includes some history of how these ideas came about. Not all the terminology about light can be explained in one chapter, so some terms are used here to start the discussion and are explained better later.

The behavior of light has much to do with how you are able to read this book at this moment. Something is coming from the page to your eye that tells you what information is on the page. You know that if there is no substantial light source—no artificial room lights and no natural sunlight—you could not read the page. Any one of several theories of light can be brought to bear at this early stage to provide explanation of how you can see a book.

1.1 THEORIES OF LIGHT

You have observed clues that the light from a source travels in straight lines from source to page and then to eye. This phenomenon is most apparent when the page has glare. Then the page func-

tions almost like a mirror because light takes special straight paths from light bulb to page to eye without bringing clear information about the dark print on the page. Since pages should not behave like mirrors, light must be used differently to read.

The simplest description or model of *light* is that it is something that travels as *rays* in straight lines. Rays are easy to visualize because we have all seen rays of light in dusty areas or in movie theaters, even as sunbeams passing through rainy clouds. Any demonstration of laser light that renders the laser beam visible shows it as a ray, or beam. Because you have not seen a ray curve, it is natural to think of rays as straight lines. However, curves and bends can occur, as when an object under water does not appear to be located where it really is. Any ray traveling from water to air and then to the eye changes direction.

Light must travel from this page to your eye in order to convey information. The path is straight, or nearly so. However, anyone located in another direction nearby also can read the book, so light must be traveling similarly to him or her. Therefore, light is traveling in very many directions from this page. This fact holds true even if the source is just a point of light, like a pinhole in an opaque lampshade. A ray of light striking the page is converted to many rays of light as it leaves. Whether you see glare depends on what angles the light reaches and leaves the page and on how slick the surface is. The detailed structure of the illuminated surface has much to do with how light is affected by the surface.

That light propagated in straight lines was stated at least 2300 years ago. That it took a path requiring the least time to travel was proposed by Hero of Alexandria in about AD 100. But what exactly is a ray? It is better to think of it in the abstract description—just a mathematical line—in contrast to the stream of fluid or tiny particles originally proposed by Greek philosophers.

A model that seems to tell what light is, is the *corpuscle model*. It was thought—in early Greece, for example—that instead of light coming from object to eye, the eye sent forth rays or particles. From reason more than evidence, members of the school of Pythagoras decided that light must be particles emitted by all objects. The general tendency in very early science was to think of everything as made of tiny particles. In this particle view, particles were understood to be something that could not be subdivided further. The corpuscle and ray models agree in that tiny particles or corpuscles can travel in straight lines. Later, in the seventeenth century, much effort went into explaining how lenses and prisms (wedges of glass) can affect the paths of corpuscles that travel through them. Isaac Newton, who demonstrated in 1672 the breaking up of "white" sunlight into colors with a prism, created a fantastic explanation in terms of corpuscles, but could not avoid a reference to waves as well.

A third model for light was developed from the notion of *waves*, familiar from water and sound waves. At least as early as 1665, light was thought of as waves in a mysterious fluid. No one could imagine waves without a medium, so the idea of a motionless ideal fluid called the ether was developed by Christian Huygens in 1690. Light waves were supposed to propagate in the ether. The breaking up of light into colors could be explained better with the wave theory. Waves in some sense can be viewed as traveling in straight lines, so the ray and wave models are compatible. Ocean waves seem to arrive in straight lines from the distant horizon to the beach. Unlike waves on water, it later was found that light does not need the ether.

In this century, the pendulum of scientific history swung again, and the *photon* as a model for light was developed. The photon is a rudimentary particle of light similar to a corpuscle, except modern photon theory is very precise about what a photon is and what it does. Photons, like any particles, even reflect from smooth surfaces just as if a ray were reflecting from a mirror. More difficult, but still feasible to explain, is how a photon can be brought into focus by a lens. There is an intimate connection between photons and light waves, although they seem to be very dissimilar models. Either can be used separately to explain most phenomena.

Another kind of model for light is based on *information*. This is a more abstract model in which the concern is not with what light is or how it does things but rather with what it carries. Light rays, waves, or photons carry information, and the amount of information can be defined precisely, depending on what is done to the light. The carrying of information is closely related to the fact that light carries energy. You cannot have one without the other.

At the moment, your reading of this page does not provide any obvious way to distinguish between ray, wave, and photon models. Later in this book, very subtle behaviors of light will be shown that will bring out these differences. The limit to the accuracy with which you can distinguish print might be evidence of wave effects. The fact that you need a minimum amount of illumination before you can see the page at all is evidence of a minimum energy for photons.

1.2 MODELS, LAWS, AND THEORIES IN SCIENCE

This section is a necessary digression to clarify some language that has been used and will be used often in this book. A ray, wave, or photon is a *model* of how light behaves. Rays tell something about what light does: it goes in straight lines. Waves tell something else: that something "waves" as light travels. Photons give still another kind of information: that light arrives in "chunks" or units. What light is, is less tangible. These are merely simplified descriptions of aspects of light. A model in science is intended to be as simple as possible, while still doing the job needed. The fact that there are three or more different models of light is a clue that light might be very much more complex. Henceforth when it is said that light "is" a wave or "is" photons, a convenient simplification has been made. A model does not constitute an explanation or a complete description of an aspect of nature.

A major activity in science is making observations about nature. They are set up to be as objective as possible, making them independent of the particular subjective prejudices of the scientist doing the observations. Many observations are made in controlled situations called *experiments*. For all the things that can happen during an event such as a flash of light, as many as possible are controlled or measured. For example, if the interest is in how the heat applied to a substance causes flashes of light (thermoluminescence), then possibly irrelevant factors such as humidity are kept constant. Heat input and light output are called *variables*, and the relation of the two is measured.

If many scientists repeat the experiments and agree on the observed behavior, then the results might gain so much acceptance they are called a *law*. A law in science is not a human decree, but rather the final and simple result of many careful observations about a particular phenomenon (Fig. 1-1). A law can be expressed in verbal or mathematical form. A natural law can be overthrown if more careful work finds a better one. A law found by scientists is not considered to be an explanation of nature, just as the law of gravitation does not tell what gravity is, only what it does. A law enables predictions of events in new ranges not originally covered by experiments, and if these events are found, the law is further respected.

To explain what light is, a *theory* is needed. A simple theory could state that light acts as a wave of a certain type. The description can be verbal or mathematical. Mathematics is very handy because laws and theories in this form have strong predictive power. Instead of suggesting, "look there and you will find that," a mathematically based theory can say clearly, "calculate such and such and look for this exact result."

A theory is more than a model in that it gives all the conditions for use of the model—for example, using electromagnetic waves for light instead of soundlike waves. A theory is more than a law because it provides some explanation of how nature works and has strong predictive power. A theory of gravity would not only enable the calculation of the strength of gravity on any given planet, but would tell what happens between planetary mass and human mass to cause an attractive force. The electromagnetic theory of light triumphed because it predicted many properties of light from earlier, simpler laws found from experiments on electric currents and magnets. No light was involved at the experimental stages!

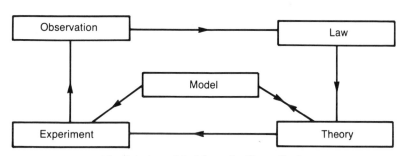

Fig. 1-1. A model of the scientific method.

1.3 SOME PROPERTIES OF LIGHT

You cannot read this page unless the light you are using has certain necessary properties. The light must have certain colors; that is, it must be visible. Moreover, a mixture of colors that makes near-white is desirable. Imagine trying to read print in green light or even red light. If the paper were colored, rather than white, then certain colors of light would render the paper as black as the print. In some ways, it might be easier to imagine photons, rather than waves, as having color. A wave is a diffuse thing whose behavior varies over time, whereas a photon is a chunk of well-defined energy and therefore color. How either of these models can have color is one of the fascinations of optics.

The light also must be bright enough, but not too bright. A better term is *intensity*, related to the energy light carries. Dim light such as starlight is entirely unsuitable for reading, yet astronomical photography makes very good use of it. At the other extreme, light can be made so intense that the page will heat up and ignite. Sunlight has this intensity at approximately the location of the planet Mercury.

Light has several other properties not apparent to the casual observer. Its *phase, coherence, polarization*, and *quantization* (all to be discussed much later) do not normally affect the process of seeing print on a page. Animals, however, can see polarization and thereby navigate with the sun as guide. If your light source were a laser (not recommended for reading) phase variations might be very apparent. Light, when it behaves like particles, must have a property akin to mass. The effect of photons bouncing from a surface is mechanically similar to tiny balls bouncing.

Light can be turned on and off rapidly. When the light is off, the rays "vanish" or the photons stop coming. The waves die out. When light is turned on, the waves must start up, but the photons do not start up—they are suddenly traveling. Light sources can flicker. The light you are using is probably flickering 120 times per second if it is artificial, and you can detect this by sudden motion. Certain rates of motion will seem to stand still or, "strobe." Beyond a certain rate, flicker appears as a steady light source due to the slowness of the eye's response. A television screen flickers 60 times per second, but you see only smooth motion as the images change.

Another property not apparent during reading is the speed of light. For a long time it was thought that light traveled infinitely fast, and early experiments were too inaccurate to prove otherwise. For example, if two people on widely separated hills flash lanterns at each other, neither person can determine any time delay between the first flash and the reception of a answering flash. Galileo, as a spinoff of observing the moons around Jupiter, realized that the light from the moons was taking a definite and rather long time to reach the Earth and estimated its speed. Olaus Romer refined the measurement in 1676. Mechanical determinations of the speed were done by Armand Fizeau in 1849 with a rotating "chopper" wheel, and by Jean Foucault in 1850 with a rotating mirror. Foucault also showed that the speed was less in water than air, contradicting the popular corpuscle theory and favoring the wave theory.

Twentieth century physicists, such as Albert Michelson in 1926, continue to strive for more accurate measurement of the speed of light. Like many pioneering experiments in physics, an approximate measurement is now a simple undergraduate laboratory exercise. The fact that the speed of light is a constant in empty space (vacuum) has implications far beyond optics. Inside a transparent material, the speed is less (for complex reasons to be shown later). Although large, the speed of light is now a basic barrier to many technological goals, such as faster computers.

If light is waves, what "waves"? It is easy to show that the air between your eye and this page is not necessary for the light to reach you. However, does light need some material—or immaterial—medium to travel, just as sound waves need the air? What, if anything, is all around us that supports the propagation of light waves so that we can see the illuminated world?

Perception is covered in detail later, but a few remarks are pertinent at this point since everything we see depends on the biology and psychology of perception. For every physical property, there is a corresponding psychological property. It will be im-

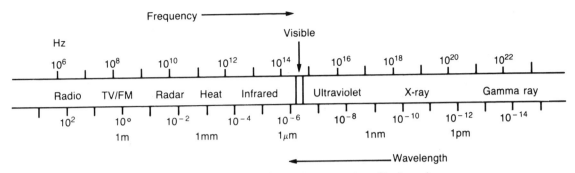

Fig. 1-2. The electromagnetic spectrum on a logarithmic scale.

portant to distinguish the brightness of light, a very subjective perception, from the objectively measurable intensity of the light. Similarly, the perception of hue for light is subjective, but the color of light can be defined and measured objectively. Other such pairs of subjective and objective variables will be introduced as needed.

The range of theories, laws, and applications covered in this book must be limited, and some limitations can be defined in terms of certain properties of light. The colors of light indicate a property of light that can be measured over a wide range, or *spectrum* (Fig. 1-2). Technically this property is defined using frequency and wavelength, which will be explained later. *Color* is a name for this variable for visible light. Optics, in contrast with sciences dealing with other parts of the spectrum, is commonly understood to deal with visible and near-visible light, a very narrow part of the available spectrum. Near visible light includes "colors" redder than red (*infrared*) and "colors" more violet than violet (*ultraviolet*). The range extends just far enough to either side of visibility that the light can still be handled by most lenses and mirrors made of the usual materials. The changes in optical properties of materials as color is changed are often crucial, however.

The principles of optics apply to all colors or wavelengths of electromagnetic waves. Radio waves are like very, very red light, and X-rays are like very, very energetic violet light. There is really no end to the spectrum. However, the study of radio waves, radar, X-rays, and beyond are different fields of physics with different applications. Henceforth, light of interest in most optics will be defined as the range from infrared to ultraviolet. Effects at very low or high temperatures, or waves/photons sufficiently energetic to attack the nuclei of atoms, are beyond the scope of optical physics.

In regard to other properties of light, there will be no limitations. Light of any intensity down to one photon per hour or less—very, very weak light—and light sources far brighter than the sun—bright enough to boil matter—can be of technological and astronomical importance.

1.4 UNDERSTANDING AND USING LIGHT

The physical theories of light are relatively complete as well as simple and profound. The interaction of light with matter is in principle understood, although practical calculations and fabrication are often very difficult. The relation of light to living organisms is less easy to understand. Light is used in fine instruments to explore the worlds of the very small, both inanimate and animate, down to the limits of the size of the waves. Light from stars and galaxies is interpreted in increasingly complex ways to determine the secrets of the universe, even its origin. Laser light is an investigatory tool of great precision that has deeply changed most sciences and technologies.

In addition to allowing the reading of this book, light has aided humans, animals, and plants in many ways by rendering the world visible. If all life were blind, vision soon would evolve because the advantages of vision are so great and detailed accurate seeing is biologically quite feasible. The way eyes provide usable images of the world to brains is very complex and just beginning to be understood. Almost all life, including bacteria and plants, are sensitive to light in some way. Plant stems find it useful to grow toward the brightest light, while roots help by growing away from it, seeking the depths of the soil. Plants have had a major effect on our planet by converting light energy to useful food—a process too complex to describe in detail in this book.

Light has much to do with technology through the information it can carry and through the precision its tiny waves/photons permit. All properties of light have been thoroughly exploited. From the first art and books to the latest holographic, laser, and optical fiber technology, light has been harnessed by human cultures. Media dominate the world through the videocamera, television receiver, and motion picture. All are special applications of optics. The science and art of photography delight many and support all modern publishing, down to the printing plates themselves. The control of light in artwork is an ever-fresh challenge to artists and a continuing delight to art lovers. Those unfortunate to lose or never have sight eventually might have reasonably accurate vision thanks to various attempts to couple light sensors to brain functions.

2

Waves

You will need an acquaintance with waves at an elementary level to understand the electromagnetic theory of light waves. In this Chapter, some basic terminology about waves is introduced, using simple observable examples of waves in materials. Water waves, waves on strings, and sound waves provide familiar examples of mechanical waves. Each property of waves will be reexamined in later chapters, so you will have several chances to become familiar with them. The notion of waves is very general, transcending the particular examples. Sophisticated properties of waves are also introduced, using familiar examples. Although some of these examples are not relevant to light waves, they should help you gain a general feel for the various properties of waves before you study light.

2.1 BASIC PROPERTIES OF WAVES

Waves are familiar in the form of waves on water and some aspects of sound waves. Waves are also readily observable on stretched strings and wires (see Fig. 2-1). Water waves display the form and motion of one type of wave. On water, the "hills" and "valleys" of the wave can be seen. Sound waves have similar properties, but you cannot ordinarily see how they work; you can only hear the results. The wave aspects of light are not at all apparent in everyday situations.

As revealed on water, a wave is a disturbance of a material called a *medium*. The disturbance might move, but the medium as a whole does not. The moving wave is said to be *traveling*. The disturbance represents energy in the wave. That this is real energy and can travel from one place to another can be seen in the example of a tidal wave. It might start at a distant place but does much damage when it reaches ships and docks.

Sound waves consist of very small disturbances in the air or other materials. A parcel of air does not actually move from a speaker to a listener. Instead, air molecules are compressed slightly by pressure from moving lips and tongue. These pressure changes move rapidly away at the speed of sound.

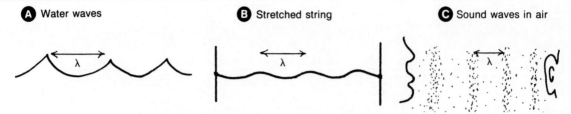

A Water waves   **B** Stretched string   **C** Sound waves in air

Fig. 2-1. Three kinds of waves. (λ is the wavelength.)

They become weaker, but otherwise their form is unchanged, so the listener can understand the speaker. At least one sort of wave—light or electromagnetic—does not need a medium in order to carry energy.

The energy carried by waves needs a continuing source, or the wave motion dies or damps out. In a large pond, a disturbance at one place might not travel to all parts of the pond before the small amount of friction in water takes away all the energy in the waves. Water waves seem to have similar forms and regular spacing. Wind can cause an indefinitely long series of moving waves, some larger than others, but all traveling at the same speed. Therefore no waves catch up with and pass others.

One type of wave, which can be seen in a small tank of water or on a string, does not move. The water or other medium bobs up and down in place, and the wave is called *standing*.

Experiment 2-1

Drop two or three stones widely separated into a still pond and observe the circular waves that develop from each impact. Watch their behavior over time. ∎

Waves begun from one place in any simple medium travel outward the same in all directions. This produces symmetrical circular shapes for water waves that result from a small disturbance (Fig. 2-2). Another kind of wave has a flat or straight form

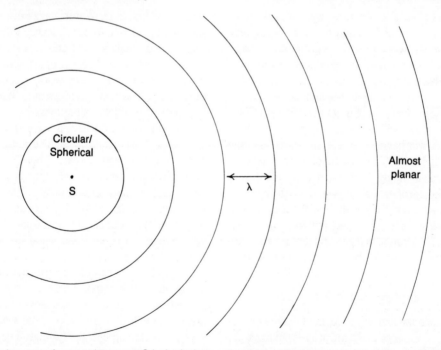

Circular/
Spherical

•
S

Almost
planar

λ

Fig. 2-2. Circular waves from a point source S (spherical waves in cross section) become almost planar at large distances.

Fig. 2-3. Interference pattern from three sets of circular waves.

and can be made in an approximate way by moving a long flat board in water. Another way to obtain such *plane* waves is to be far from the source. Ocean waves come in relatively flat because their cause is usually far away. Sound waves travel in all directions in three dimensions, so their form is approximately spherical. The cross section of a spherical wave is a circle, which is what must be shown in illustrations.

Observations of water waves show that such waves coming from several sources can pass through each other as they travel. Although the waves intermingle, they do not seem to mess each other up permanently (Fig. 2-3). This combining of waves is called *interference*, but the waves do not literally interfere with each other. Rather they temporarily combine. Of interest in interference is the formation of stationary patterns where the waves interact.

Experiment 2-2

If you have a small tank or pond of water, float or place a barrier across the middle. Start circular waves moving toward the barrier. Where do they go? Send plane waves at various angles using a flat board. Change the barrier to one with a small opening, so that there is a continuous water path from one side to the other. Start a circular disturbance on one side and observe the resulting waves on the other side of the barrier. ■

Water and other waves reflect from certain kinds of barriers in the medium. The direction changes in the special, simple way familiar from mirrors, and the wave keeps going as before (Fig. 2-4). If the barrier has a hole, a new and less familiar effect occurs. The waves arriving at the small opening become the cause of a new set of circular waves (Fig. 2-5). A small opening acts as a local disturbance. The spread of waves after being constrained by a small opening is called *diffraction*. Sound and light waves undergo interference, diffraction, and reflection in ways similar to water waves, although the detailed mechanisms are different.

Experiment 2-3

Tie a rope between two posts, stretching it tight. Shake one post in a regular manner and ob-

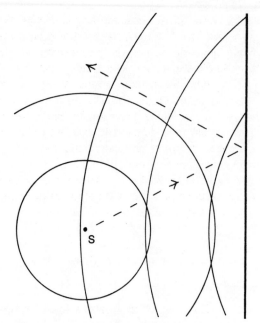

Fig. 2-4. Reflection of circular waves from a smooth surface (in cross section).

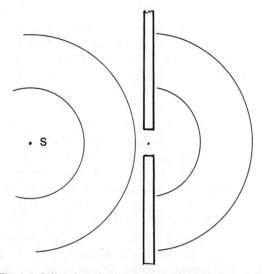

Fig. 2-5. Diffraction of circular waves encountering a hole in an obstacle (in cross section).

serve the waves on the rope. Pull down the rope with a finger near one end and observe the resulting wave. Can the same effect be made sideways? ■

There are many ways to start waves, depending on the medium provided and the type of wave wanted. A stretched string, wire, or rope can have traveling waves that start from a disturbance at one end and progress to the other end. If the disturbance is a regular vibration at certain rates, standing waves can be established. Then the string moves up and down without the "hills" and "valleys" moving along the string. Hills become valleys become hills, over and over at a regular rate.

A form of wave called a *pulse* is started on the stretched string when a single pull or jerk is made at one place on the string. Two pulses then travel rapidly along the string (slower on a heavy rope) in opposite directions. When a pulse reaches an end, it changes direction and returns. This process continues for several trips while a complex rearrangement occurs. The result is a standing wave with a smoother shape. Thus are traveling and standing waves connected. A standing wave on a string is also called a *vibration*. It is also possible to start a single pulse (solitary wave or *soliton*) moving on water and in air and other media.

2.2 MORE ADVANCED PROPERTIES OF WAVES

A stretched string can vibrate in different directions: up and down, sideways, or diagonally. In every case, the material of the string moves sideways, but wave motion is along the string. Such a wave is *transverse*. There is another way the material can vibrate: back and forth along the direction of the wave. A sound wave is such a *longitudinal* wave. Water waves seem to be transverse, but the situation is actually more complex.

Experiment 2-4

Cut a slot in cardboard and pass the stretched string through it without touching. Make the string vibrate vertically and test the results of holding the slot horizontally and vertically. ■

If a string can vibrate up or down, or back and forth, the wave is said to have *polarization*. Like a string, light can vibrate in only one direction at a time. (Later it will be seen that a bunch of light waves can vibrate in all directions at once and be unpolarized.) Polarization is one of the most difficult

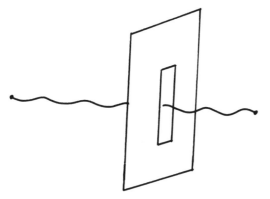

Fig. 2-6. Transverse waves are polarized and can pass through a slot.

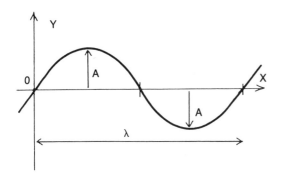

Fig. 2-7. Graph of part of a sinusoidal wave.

aspects of light to understand. The filtering of polarization is readily demonstrated with a vibrating string (Fig. 2-6). If the string passes through a slot in a cardboard without touching, there is only one way the string can vibrate without having its energy damped out from touching the edges of the slot. Whichever way the slot is oriented is the only permitted polarization of the wave.

The more energetically you shake a string or its fastenings, the larger the transverse motions of the string. The shape might appear messy at first, but eventually it can settle down to a common form. If the resulting wave has a simple, regular, and repetitive form, the size of its motion can be described by one characteristic, or *parameter*, the *amplitude*. If the amplitude is small, the wave might have a very smooth shape, called *sinusoidal* (Fig. 2-7). Waves on strings are difficult to observe unless the string is a long, heavy rope vibrating slowly

in place (standing). Water waves appear choppy, but very small disturbances on water should have the smooth sinusoidal shape. The pressure changes in sound waves also vary smoothly in the same way.

The repetitive nature of most waves is described in terms of *wavelength*, *period*, and *frequency*. The wavelength of a low-amplitude water wave is readily seen to be the distance between consecutive hills or crests, the same as the distance between consecutive valleys or troughs. Wavelength (symbol λ called lambda) is also the distance between two consecutive peaks on a vibrating string or between any two identical parts of the wave. If you could see a sound wave of one pitch moving in air, there would be a regular distance, λ, between consecutive regions of high (or low) pressure.

Experiment 2-5

Tie one end of a long rope to a stationary object. Hold the other end and practice shaking it up and down at various rates until you hit a special rate. Then the rope will settle into one or more smooth standing waves. ■

There is a special relation between the wavelength of waves that can be made to stand on a string or rope and the rate at which it must be shaken. The time between shakes, usually quite short, is the *period* of the standing wave. If you have heard, or can make, very low pitched sounds (lower than the bottom note on a piano), you might be able to hear the individual crests of the sound waves. Generally the ear cannot distinguish times smaller than about 1/20 second. The period of a water wave is simply the time between the arrivals of two consecutive waves. It should be measured in deep water.

The *frequency* of a wave is related simply, but inversely, to the period. The frequency is the rate at which something occurs. If a wave repeats at 0.10-second intervals, then it repeats at the rate of 10 times per second. This is also called 10 Hz (hertz).

Each cycle of a wave consists of a crest, a trough, and two instants of no displacement of the medium. What part of the cycle the wave is considered to start at, or be detected at, is called the *phase*. Suppose in a certain reference system, waves

Fig. 2-8. Two possible phase relations between two identical waves.

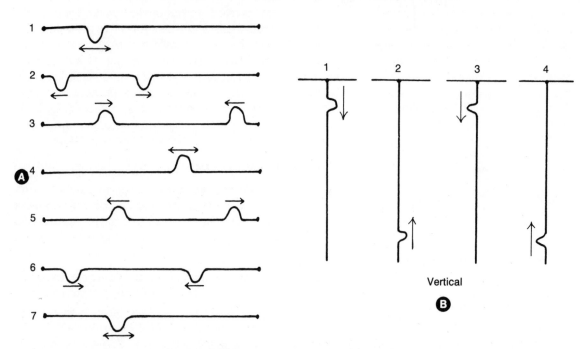

Fig. 2-9. Reflections of pulses on a stretched string.

are defined, or caused, to start at zero displacement and build up, and the waves are detected at the crests. Then detection has occurred at one-fourth cycle, or at a phase of 90°. If two sets of waves are intended to be matched, then the simplest attribute to match is the phase. If crests are put together, the waves differ by zero phase or are *in phase* (Fig. 2-8A). If crests of one are matched with troughs of another, the phase differs by one-half cycle or 180°, and the waves are said to be completely *out of phase* (Fig. 2-8B).

Reflection occurs for waves on a stretched string as well as for water and sound waves. The explanation is unusually simple. Consider the single wave or pulse started on a stretched string fixed at both ends (Fig. 2-9A). Two downward-deflected bits

of string each travel to the ends, where they tug on the fastenings. An old law of physics from Isaac Newton states that at any point, the action in one direction is balanced by the reaction in the opposite direction, provided the point does not move. Because the fastening does not move, it cannot help but tug on the string in the opposite or upward direction. Consequently, bits of string start moving the other way as shown. After another reflection, pulses are moving back with the original orientation and motion. They coalesce momentarily as they pass each other.

Experiment 2-6

Stretch a rope between two posts and observe the orientation of the pulse that returns after pulling sharply down on the rope near one end. If you hold one end of the rope, your hand is the fastening. Can you feel the returning pulse jerk your hand? Now hang a long rope vertically, leaving the lower end free in the air. Start a small pulse near the held end and observe its orientation after reflection. ■

A different case of reflection is obtained when a rope is hung with one end free (Fig. 2-9B). There is no action-reaction at the free end, so the pulse (just one is shown to avoid confusion) comes back on the same side that it went down. At the fastened end, the force from the pulse produces a reaction that pulls the rope in the opposite direction. This time, to obtain a pulse moving back the same way in the same orientation, four reflections are needed.

There is a close relationship among reflection, period, wavelength, traveling waves, and standing waves. On a string or in any medium, the time for a pulse or other traveling wave to repeat exactly what it was doing before is its period. After repeated reflections, a pulse of whatever shape interferes with itself in such a way as to form a standing sinusoidal wave. Often the wave on a string has the shape of an arc and a wavelength equal to twice the length of the string. These results apply to any bounded situation. Sound waves in a solidly built, closed room can be standing at certain periods and wavelengths, as anyone singing in the shower is aware of at certain pitches. The application to light is important to modern technologies, such as the laser.

Thus far, in the discussion of waves, solid boundaries have been used with stretched strings and other wave media. What if one boundary is soft and somewhat flexible? Consider a two-component string consisting of a cord tied to a rope, stretched in series and very long so that there need be no concern over the fastenings (Fig. 2-10A). Let the light end be attached to a *wave generator*, a rapidly moving arm that shakes the string in a regular way. This is a convenient way to test waves on strings and will start a wave traveling along the lightweight part of the string. When the wave encounters the heavy rope, it considers the boundary to be semirigid. Therefore, some of the wave will reflect as usual. The end of the rope can be shaken a bit, so some of the wave continues on. However, that is not the end of the story. Strings with different weights carry waves at different speeds. It seems sensible that a heavy rope carries waves slowly, other things being equal. A massive rope is hard to shake quickly.

Meanwhile, the wave generator keeps on shaking at the same rate, producing waves of a certain wavelength on the light string. The same rate must continue on the heavy rope, but because the waves move slower, they pile up in such a way as to have shorter wavelength. The change in string weight caused a change in speed and wavelength. This is a special case of *refraction*, the change in speed of a wave as it enters a new medium. A heavy rope is a different medium than a light string. Refraction can be seen for water waves approaching a beach (Fig. 2-10B). In shallow water, the natural speed of waves is less because there is more drag and so forth. Therefore the distance between crests

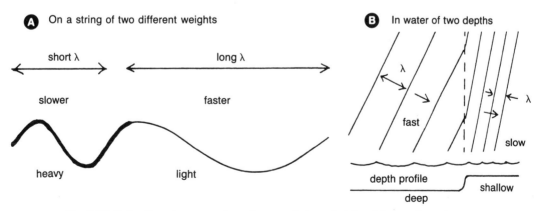

Fig. 2-10. Refraction of a wave where the speed characteristic of the medium changes.

decreases, while the waves arrive at a constant rate. If the wave arrives at an angle, the wave direction must bend as the wave slows (as will be shown later). The same effect occurs with light entering a glass prism, although why the colors separate is not yet as clear.

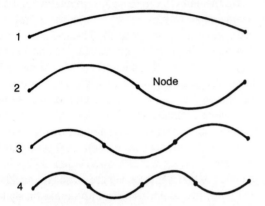

Fig. 2-11. Different modes of vibration on a string with both ends fixed. The fundamental has no nodes, the second harmonic one, and so forth.

Experiment 2-7

Try the preceding rope-string experiment for real. Omit those parts needing a mechanical generator if you cannot rig one. ■

More properties of waves can be found when a generator shakes a string. A simple, lightweight, stretched string possesses certain characteristic frequencies of shaking at which it will have standing waves. As the frequency increases in steps, the wave shapes are progressively changed as shown in Fig. 2-11, with more and more bends in the smooth shape. The lowest frequency gives the *fundamental* shape (a half-sine). The next frequency gives a new *mode*, called the *second harmonic*. A frequency three times higher than the fundamental gives the *third harmonic*. The fixed ends, of course, do not move, and there are certain other points (*nodes*) that also do not move. This can be seen by considering one of the shapes in snapshots over time, as shown in Fig. 2-12. Sometimes the hills are one way, sometimes the other. The pattern repeats

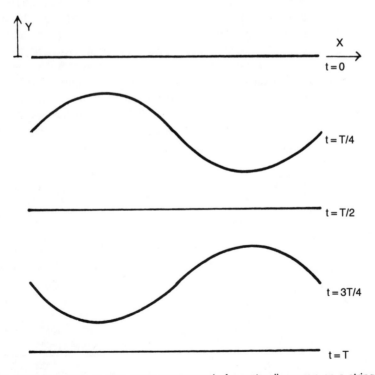

Fig. 2-12. Snapshots in time every quarter cycle for a standing wave on a string.

in exactly one period. Usually the motion is too fast to see, and it appears as a blur with visible nodes. Any driving frequencies other than these special ones cause an erratic and damped response. Repetitive behavior over distance and in time are closely related and vary in the same smooth way.

There are other possible media for waves. Waves on stretched sheets or membranes, for example, are very complex. These media are useful in physics, technology, and many other fields including music. Rods or bars of wood, steel, and other materials can carry waves and pulses which are best thought of as acoustic rather than sound waves. Imagine someone striking a railroad track rail a kilometer away. The clang will arrive at your ear through the rail much faster than the sound through the air. Water can carry acoustic waves in it, as well as surface water waves. In some media, the wave shape cannot be sinusoidal and/or the frequencies of the modes do not have simple integral relations.

Membranes, bells, bars, and other shapes have natural frequencies (modes), just as stretched strings and closed rooms have. The modes are always evoked when the disturbance is very brief, as when hitting it with a hammer. A natural frequency is aided if the material is free to vibrate, and it might not appear at all if the material is constrained in the wrong places. Earthquakes are relatively brief, intense disturbances in the earth that serve as the source of several kinds of long-wavelength waves in the solid earth.

2.3 QUANTITATIVE DESCRIPTION OF WAVES

Simple mathematics permits useful descriptions of many properties of waves. To make the mathematics serve the physicist and engineer, it must be possible to work with both symbols and numbers. This is the quantitative approach to studying and utilizing nature.

Numerical values for waves are best measured in the near universal *SI* (*Systeme Internationale*) unit system. In this system, lengths are measured in meters (about two-thirds of a person's height), time in seconds, and mass in kilograms. In this book, there will be little need for other special units. The speed of waves is measured in meters per second. Large multiples of units (thousands, millions, and so forth) and tiny subdivisions of units (10^{-3}, 10^{-6}, etc.) are denoted by prefixes to the units. The prefixes and associated powers of ten are summarized in Table 2-1.

If you have studied trigonometry, you can immediately use a special and simple mathematical description of a wave. A sinusoidal wave thought of as a standing wave on an ideal string was shown in Fig. 2-7. The string is along the x-axis, and the transverse displacement of the string is graphed on the y-axis. If the string is fixed at the origin O, causing a node there, then the wave is described along the string by:

$$y = A \sin (2\pi x/\lambda)$$

Table 2-1. Powers of Ten Notation.

Power	Prefix	Symbol	Old Name	Explicit Form
10^{18}	exa-	E	quintillion	1000000000000000000
10^{15}	peta-	P	quadrillion	1000000000000000
10^{12}	tera-	T	trillion	1000000000000
10^{9}	giga-	G	billion	1000000000
10^{6}	mega-	M	million	1000000
10^{3}	kilo-	k	thousand	1000
10^{0}	...	...	unit	1
10^{-3}	milli-	m	thousandth	.001
10^{-6}	micro-	μ	millionth	.000001
10^{-9}	nano-	n	billionth	.000000001
10^{-12}	pico-	p	trillionth	.000000000001
10^{-15}	femto-	f	(etc.)	.000000000000001
10^{-18}	atto-	a		.000000000000000001

A is the amplitude of the wave, and λ (lambda) is the wavelength, each having some fixed value. The *sin* function is shorthand for converting the numbers represented by the *argument* $2\pi x/\lambda$ written after *sin* into another number between -1 and $+1$. The *sin* is an abbreviation for sine and is the source of the term sinusoidal, which refers to the shape of the function as graphed. The *sin* function is periodic in its argument, which means that as x increases steadily along the string, the *sin* function varies smoothly between -1 and $+1$ in a regular manner. To obtain a wave of any desired size, the whole thing is multiplied by A. At certain values of $2\pi x/\lambda$, such as 0, π ($180°$), 2π ($360°$), and so forth, the *sin* function is zero, producing nodes.

The wave varies as smoothly in time as it does in space. For any point x on a vibrating string, the displacement of that point over time is given by:

$$y = y_0 \sin (2\pi t/T)$$

Here T is the period of the wave. y_0 is the maximum amplitude the string can have at position x as determined by the previous graph of the wave.

Period and frequency (f) are inversely related by:

$$f = 1/T$$

Students of trigonometry can readily see how wave interference of sinusoidal waves can result in new sinusoidal waves. The sum of two sine waves at a position is simply another sine wave, according to a theorem in trigonometry. The arguments combine in a special way that will be useful later.

2.4 THE BASIC PHYSICS OF WAVES

In many cases of small vibration—such as water waves, stretched strings, sound waves, and light—a special and simple physical law determines the wave behavior. Because waves involve motion of mass, Newton's law in the simplified form $F = ma$ applies. Every case must have a *restoring force* F that pulls any displaced mass back to an equilibrium position. For water waves, a combination of gravity and surface tension tend to smooth the surface of the water. For stretched strings, the increased tension in the string tries to restore displaced pieces to the rest position. For sound waves, the increased pressure of parcels of air tends to bring the air back to uniform pressure and fill in regions deficient in air. For light, there are complex restoring forces. In all these cases, the restoring force directly proportional to the displacement, which inevitably results in a sinusoidal wave.

The physicist can set up an equation that relates restoring force to motion of the medium. In simple cases as described, this equation can be solved for what is called an *analytic solution*, a mathematical relation such as those shown that tells what the wave is doing at all places and times. In complex cases, the only solutions might be numerical approximations best found on a computer. The result of solving the equation is to find unexpected results in addition to the sinusoidal motion. Of great interest is a calculation of the wave speed in terms of parameters for the particular situation.

There is a general relation between wavelength, frequency, and wave speed:

$$v = f \lambda$$

(That there are two kinds of wave speed is left to advanced physics students.)

For large water waves (wavelength larger than about one meter), the wave speed in deep water is given approximately as:

$$v = \sqrt{g\lambda/2\pi}$$

where g is the acceleration of gravity: 9.8 meters per second squared. In this case, the speed of the wave depends on the wavelength. Ocean waves, when made four times longer travel twice as fast. A one-meter-long wave travels about 0.6 meters per second. Note that the units of $\sqrt{g\lambda}$ are the same as the units of speed squared; thus the equation balances in terms of physical units. Closely spaced and, therefore, small ocean waves have their behavior determined more by surface tension than by gravity, and the speed relation is more complex.

For a stretched string the speed of waves is given as:

$$v = \sqrt{F/\varrho}$$

where F is the tension in the string and ϱ (rho) is the mass density per unit length. F is measured in newtons, a unit of force in SI units.

For a sound wave in air the speed is given approximately as:

$$v = \sqrt{\gamma p/\varrho}$$

where p is the ambient air pressure, ϱ is now the density of air, and ω is a constant for air, approximately 5/3. For rough estimates, the speed is about 330 meters per second, but it depends on temperature, altitude, humidity, and other factors.

For light the wave speed depends on the material the light travels in. It also depends slightly on the wavelength or frequency. Blue light travels slightly slower than red in most materials but at the same speed in vacuum.

Experiment 2-8

To learn more about how water waves really act, put a small floating object on water and make waves. You might need to go to a lake where waves are large and slow. As a wave passes by, observe the motion of the object. ∎

As with other waves, water waves do not transport water, just energy. A floating indicator on water will show that parcels of water cycle in place, rotating forward, down, back, and up. Thus, water waves have both transverse and longitudinal displacements combined. It is not surprising that the wave speed depends on wavelength, unlike the simple transverse waves of strings and longitudinal sound waves.

3

Electromagnetism and Light

In this key chapter, light waves are described in all their theoretical glory. The mathematical aspects are optional in the reading, but some mathematical symbolism is used for convenience. Simple mathematical relations are provided where pertinent. More advanced ones are reserved for the last section of the chapter. If you have only little technical interest, you can browse here and there for the basic principles; then proceed to the optical applications in the next chapter.

The discussion must start deep in the experimental foundations of physics in the nineteenth century. Each set of original experiments that was found consistent resulted in a new law in the physics of electricity or magnetism. Law by law, a total picture was constructed of how electrical charges can cause waves. Expressing the laws of electricity and magnetism without mathematics is difficult, but is attempted here. You will see that electromagnetic waves are a phenomenon described by the laws of electricity and magnetism, and that they are the same as light waves.

3.1 ELECTRIC AND MAGNETIC FIELDS

It is assumed that you have a slight acquaintance with the idea of *electric charge*. The units of charge are *coulombs*. Charge comes in two kinds: positive, and negative. All matter contains both kinds of charges in equal amounts, so matter appears to be electrically neutral. For this book, it is sufficient for you to know that *electrons* are the very small negative charges that are somewhat loosely bound to the atoms and molecules of matter. Sometimes rubbing two materials is sufficient to cause a small separation of charge. Electrons are transferred to one material to render it negative, and the other material is left positive because of the loss of electrons. When the separated electrons try to return to the positive material, electric sparks might be seen. When charges move in space or in a wire, they constitute an electric *current*, another concept needed to understand fields. The units of current are *amperes*, and 1 ampere equals the flow of 1 coulomb per second.

Electric charges and currents have been shown

experimentally to influence the space around them. This is a way that one charge can "know" it is attracted to or repelled by another charge. It is also the way that a compass needle can "know" to respond to other magnets. Both charges and currents have *fields* around them, which announce to all space that charges and currents are present. Such fields are not visible to us; only their effects are felt as forces on charges or currents. A popular representation of a field consists of imaginary lines that follow the direction of the forces the fields can cause on charged matter (Fig. 3-1). The field around stationary charges is purely an electric field (symbol E). The field around moving charges is a mixture of electric and magnetic (symbol B). The field around a wire carrying a current is purely magnetic because the wire is electrically neutral whether or not the charges in it are moving.

Over 2,000 years ago, several cultures knew that matter could affect itself in mysterious ways that were not visible and that seemed to travel through the air. The possibilities of charged and magnetized matter were known from natural materials, such as amber and lodestone, although the causes were unknown. Much later, in the eighteenth century, the idea of a field as an influence in space gained acceptance. Earlier explanations of the forces between charges, and between magnets, was "action at a distance;" this was not an especially clear explanation. Between 1785 to 1789, Charles Coulomb found the law of attraction and repulsion between charges. The force between charges was found to vary according to the inverse square of the distance between them.

The relation between charges and the electric force they can exert includes a constant called the *permittivity* ϵ (epsilon). In a vacuum it has the value $\epsilon_0 = 8.85\ (10)^{-12}$ in SI units. The strength of the E field around a charge determines the force that the field can exert on another charge, so the strength is measured in units of newtons per coulomb. In some materials (dielectrics), the atoms or molecules interact with an E field so as to increase its effective strength, and ϵ is larger.

Likewise, the relation between a current and the magnetic force it can exert on a charge includes a constant called the *permeability* μ (mu). In a vacuum, it has the value $1.26\ (10)^{-6}$ in SI units. The strength of the B field is measured in webers per square meter, and determines the force it can exert on a moving charge. If the charge has zero speed, it feels no magnetic force. In magnetic materials, the atoms interact with a B field so as to increase its effective strength. Then μ is effectively larger.

In 1820, Hans Oersted showed that electric currents—flowing charges—interacted with magnetized objects (Fig. 3-2A), thus making the first link between electric and magnetic phenomena. He also showed that electric and magnetic influences had a *rotary*-type relationship—one would make the other

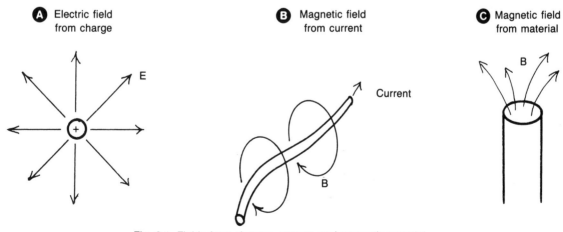

Fig. 3-1. Fields from charges, current, and magnetic material.

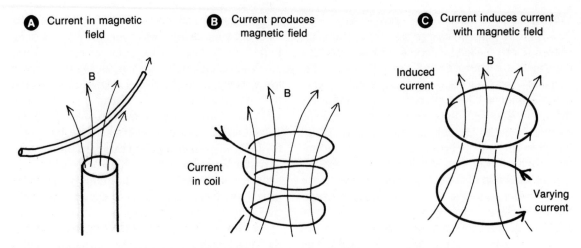

B

Induced current

B

Current in coil

B

Varying current

Fig. 3-2. Interactions of currents (moving charges) and magnetic fields.

tend to rotate, not just move in a straight line. André Ampère then found the complementary effect that current moving in circles in a coil of wire produced a relatively simple and straight magnetic field such as found from bar magnets (Fig. 3-2B). Atoms were soon suspected to consist of tiny currents (later, circulating electrons) whose magnetic fields were felt on a large scale for certain materials, such as magnetized iron.

In 1831, Michael Faraday demonstrated that a changing electrical current could induce, at a distance, another current in another coil of wire (Fig. 3-2C). The intervening influence was a magnetic field that varied due to the first current. The relation between varying magnetic field and induced current has been called *Faraday's law*. Faraday also set the stage for exploring the relation of electromagnetism and light when he showed that light was affected by a magnetic field. Other scientists, such as Charles Wheatstone and Armand Fizeau, found other connections when they measured the speed of electric current in wires and found values comparable to the speed of light.

A generalization of Coulomb's law for electric forces was formulated as Gauss's law (named after Karl Gauss). The quantity of electric field originating from charge and passing through any imaginary closed surface around the charge is proportional to the net amount of charge. The quantity is found by multiplying the field strength by the area of the sur-

face and is the total number of imaginary field lines—called the flux of the field—passing through the surface (shown in cross section in Fig. 3-3). The field and flux are zero if there is no charge enclosed. It is no accident that the area of a sphere around a charge increases as the square of its radius, just as the sizes of the electric force and field decrease with the square of the distance. Thus the total flux around a charge is constant regardless of distance.

Magnetic fields were found to behave differently. No single-pole sources of magnetic field have

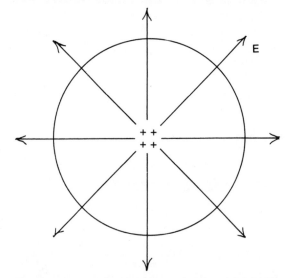

Fig. 3-3. Gauss's law governing the field through a closed spherical surface around charges, shown in cross section.

ever been found, so the net amount of magnetic flux going through any closed surface is always zero. Magnetic field lines cannot originate on anything. They must be closed loops, in contrast to electric field lines, which always start on positive charges and end on negative ones. Magnetic field lines do not originate on the north pole of a magnet and end on the south pole, but pass continuously through the magnetic material. The relation of magnetic fields to currents was well established by André Ampère in 1820 to 1825.

The result of Faraday's work was the formulation of a relation between the time rate of change of the magnetic flux passing through a bounded area (Fig. 3-4A) and a net electric field circulating around the loop. The loop could be wire forming a closed circuit, in which case a current would flow. But this was not necessary. The electric field would be formed in space as closed loops of field encircling the magnetic field at right angles as shown in perspective. This kind of electric field has no charges as sources.

Just as there is a pair of laws about the flux going through a closed surface, there might be also a pair governing flux through an open, bounded surface. A counterpart to Faraday's law was needed to preserve some symmetry in the laws. The candidate to look at was *Ampère's law*, which states

that the net magnetic field circulating around a loop is proportional to the total current through the area of the loop. It was suspected, and later found through experiments, that a changing electric field, such as that between two charged plates, causes loops of magnetic field encircling the electric field lines (Fig. 3-4B). The verification was quite late—in 1929 by M.R. van Cauweberghe. Now, the rate of change of the electric flux passing through a bounded area could be related to the net magnetic field around the loop. The symmetry between electric and magnetic fields was made complete, at least when no loose charges or currents are present. Each kind of field, when changed, could induce the other at right angles.

You can see from Fig. 3-4A that a uniform field passing by a loop will have no net result. What counts is the size of the field at each point around the loop, as compared to the direction of flow in the loop. The product of uniform field strength and distance along the loop averages to nothing around the closed loop. In some places, the field and loop are in similar directions, and in other places they are opposite.

James Maxwell corrected Ampère's law theoretically in 1865 and brought together the four laws in mathematical form to create a unified theory of electricity and magnetism, expressed in mathematical equations (later named Maxwell's equations).

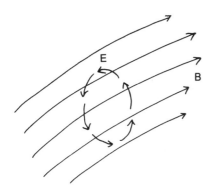

A Varying magnetic field
induces electric field
(Faraday's law)

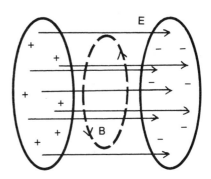

B Varying electric field
induces magnetic field
(Ampère's law)

Fig. 3-4. One field can produce the other.

The equations related in an almost symmetric way each field to the other in regard to changes in space and in time. It was possible to solve the equations under certain conditions to find electromagnetic fields—interlocked combinations of electric and magnetic fields. One simplification was to assume empty space so that no charges and currents were present. Then the equations are completely symmetrical with regard to E and B. No air, ether, or any other substance was needed, yet this mathematical theory had solutions that depicted traveling waves. Moreover, the theory predicted that the speed of the waves could be calculated from constants already measured in laboratory work.

3.2 ELECTROMAGNETIC WAVES

The simplest form of electromagnetic wave is diagrammed in Fig. 3-5. An electric field E varies sinusoidally in one orientation in space while a magnetic field B varies sinusoidally at right angles. The wave travels in a third direction at right angles to both. Thus this wave is transverse. The electric and magnetic fields are related moment by moment. They grow and shrink together. As each varies, it induces the other. The electromagnetic wave is a self-sustaining system. In empty space with negligible charges around to take away energy, the wave can travel nearly forever, like the light received from quasars 10 billion light years away. Another way to depict the fields in an electromagnetic wave is also shown at certain times (when the fields are maximum). The fields are uniform and orthogonal in planes perpendicular to the direction of travel. Periodically the fields diminish to zero then grow again but reversed.

There are many properties of this type of wave to explore. The orientation of E and B have a special relation that uses the right-hand rule. If the thumb points in the direction of E as in Fig. 3-6B, and the second finger is turned to point in the direction of B at the same time, then the third finger held at right angles to the other two will point in the direction of travel. Three-dimensional coordinate axes as represented in perspective in Fig. 3-6A are often

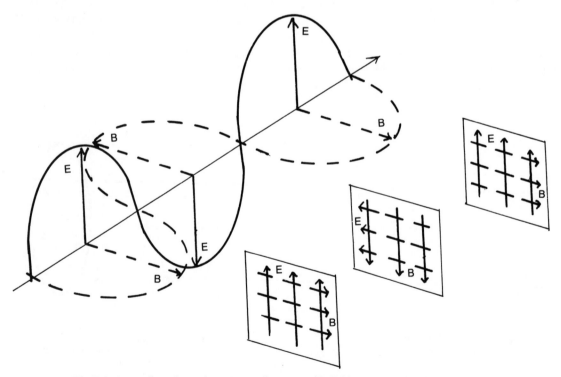

Fig. 3-5. A traveling plane electromagnetic wave, with field structure shown two ways.

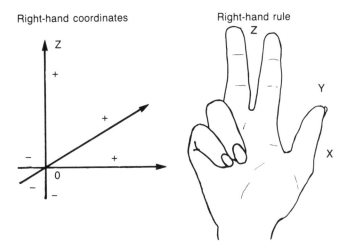

Fig. 3-6. The right-handed, three-dimensional coordinate system.

used to keep track of, and visualize, three-dimensional phenomena. By convention the x-axis is aligned with the thumb, the y-axis with the second finger, and the z-axis with the third finger, understood to point into the paper.

Visualizing how this sort of wave can occur as a spherical wave, radiating in all directions from a source, or as a plane wave, filling all space with sheets of fields, may be difficult. Plane waves result from being very far from a source, so that the spherical waves are almost flat on a small scale. As was shown in Fig. 3-5, at certain places (nodes) the field is zero everywhere in the plane. At other places each field has a certain value that holds everywhere in the plane but that varies as the wave travels. This is not realistic because no field can fill all space. Real electromagnetic waves must have boundaries. At any time, however, the strengths of the two fields are related by $E = cB$. The two fields are always interdependent. Electromagnetic waves can also be standing waves, for example in the cavity that is part of a radar tube. Then the fields oscillate back and forth in time without moving. Waves enclosed in conducting cavities have nodes at the metal surfaces and perhaps more nodes in the interior.

Where is the energy in the electromagnetic field? The energy of a wave must be described in terms of an energy density, the amount of energy in a unit volume. A theoretical wave could have energy spread through all space, giving a total energy that is infinite—and unreal. The energy density of an electric or magnetic field has been shown to be proportional to the square of the strength of the field. As the wave propagates, the energy is shuttled back and forth from electric to magnetic, with half being in each form on the average.

A traveling wave carries energy at a certain rate. The energy that passes in a unit of time is called *power*. It is most useful to consider the concept of *intensity*, which tells how much energy is carried through a unit of area in a unit of time (Fig. 3-7). Intensity can be calculated from the field thus:

$$I = c \, \epsilon_0 \, E^2/2$$

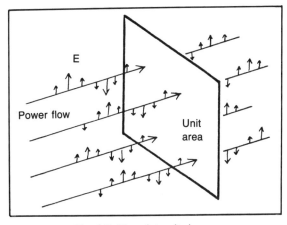

Fig. 3-7. Wave intensity is power flowing through a unit of area.

23

The square of the field helps wipe out any effect of the direction of the field on its energy or intensity. This is also called a flux. Many other terms refer to the energy of a wave, but this book will use as few as possible.

Electromagnetic waves can apply weak forces to objects they strike. A light wave absorbed by a surface tends to move the surface. The same light wave striking a reflecting surface, while depositing no energy, will impart twice as much motion to the surface. These effects are so weak they must be observed in vacuum with strong light, but the idea of shining sun or laser light on a huge reflecting surface to sail through space is a feasible prospect.

The electromagnetic wave is polarized and has wavelength, frequency, and phase. E and B each separately have polarization directions that are always at right angles. In most cases, they vary together or in phase. If c is the speed of light, then the speed, wavelength, and frequency are related by $c = \lambda f$. The speed inside materials was predicted as well as measured to be reduced by a certain amount, different for each material and depending slightly on the wavelength as well. The modern value of c in vacuum, in round numbers, is 300 million meters per second, more conveniently written as $3.00(10)^8$ m/s (to three-digit accuracy).

Producing electromagnetic waves was soon discovered both theoretically and experimentally. George FitzGerald in 1883 showed that varying an electric current would produce a similarly varying electromagnetic wave. The faster it is varied, the more effective and powerful the waves produced. The important aspect is that the current consists of moving charges. It is not sufficient for the charge to move at a steady speed. That would produce a steady magnetic field. The charge must accelerate—speed up and slow down, preferably oscillating back and forth or orbiting in a circle. The front of the wave moves out and away because each change in the source current creates another wave behind the previous one. To change an electromagnetic wave that is some distance away takes time, since each change back at the source requires time to propagate along the wave direction (at or near the speed of light). A wave received at a distance is said to be "retarded" from what is going on at the transmitter.

Heinrich Hertz made the first electromagnetic wave detector in 1886, a simple wire loop with a tiny spark gap breaking the circuit. When electric discharge occurred nearby, a tiny spark jumped the gap in the isolated circuit. Hertz also showed that invisible electromagnetic waves of low frequency (radio range) reflected like light and refracted in prisms made of tar. Soon Hertz and others were developing long distance communication via radio waves.

3.3 ELECTROMAGNETIC WAVES AND LIGHT IN MATTER

Although simple electromagnetic waves are found to propagate in space, the same wave model can be used in matter under certain conditions. All matter is composed of atoms, which in turn are each composed of light electrons bound to a heavy positive nucleus. Matter tends to be electrically neutral because the forces between charges are very strong. Because every atom is surrounded by complex electric and magnetic fields, it might seem at first impossible to describe an electromagnetic wave moving through matter. It might even seem impossible for an electromagnetic wave to make any progress through matter with all the fields in the way messing it up.

Nature provides some astonishing simplifications in matter. Light does pass freely through glass and clear plastic as well as air. Materials that do not conduct electric current well and therefore permit electromagnetic waves to propagate are called *dielectrics*. The molecules in dielectrics hold tightly to their charges, but it is possible to separate the charge slightly within each molecule. This effect is also called "polarization," but has little to do with polarization as the orientation of light waves.

Somehow (to be discussed later) the electromagnetic wave passes from atom to atom in a dielectric, encountering much interference yet keeping its original direction and frequency and losing little energy. The simplest bulk property of matter that pertains to electromagnetic waves is its index of refraction n. This quantity tells how much the wave slows down in the material. The speed of light is reduced to c/n. The value of n is somewhat different for waves of different wavelengths. Air has $n = 1.0003$, not much different from $n = 1$ for vacuum. Most glasses have values of n near 1.5. Since frequency cannot change, wavelength also changes in matter. The new wavelength is shortened to λ/n since n is always greater than one in materials.

Many materials are opaque. Light does not pass through and therefore is *absorbed* rather than *transmitted*. The energy in the wave is converted to heat in the material. In this case there is no real index of refraction. When the broad spectrum of electromagnetic waves is considered, most materials are rather selective. Radio waves might pass through while light is absorbed, or vice versa. At this point, to keep the discussion as pertinent to optics as possible, the general consideration of electromagnetic waves will be restricted to consideration of light waves, including near-infrared, visible, and near-ultraviolet. Glass is nearly transparent to near-infrared and visible light and nearly opaque to ultraviolet, which it absorbs.

3.4 REFLECTION AND REFRACTION

Reflection and refraction are very basic phenomena in both the theory and applications of optics. The law of reflection was deduced very early (ancient Greece) from observation. The law of refraction (Snell's law) was found from experiment by Willebrod Snell in 1621. It was also deduced from the corpuscular theory of light by René Descartes in 1637. The law of reflection simply states that light reflects at the same angle at which it impinges. The law of refraction gives the amount by which a light ray is bent as it passes from one clear substance to another. Air or vacuum can be one of the substances.

The laws of reflection and refraction also follow directly from Maxwell's equations for any electromagnetic wave, and that is the theoretical route taken here but without using mathematics. The process involves consideration of two regions with an interface or *boundary*, as shown in cross section in Fig. 3-8. Light is *incident* (I) in the first region and will reflect (R) into that same region as well as be transmitted (T) into the second region. All that is required is that the indices of refraction be differ-

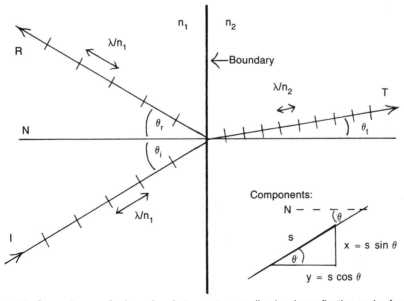

Fig. 3-8. General case of a boundary between two media showing reflection and refraction.

ent in the two regions (they are assigned the general values n_1 and n_2). The angles of incidence, reflection, and transmittance are all measured from the perpendicular or *normal* to the boundary. These angles may seem contrary to a common sense approach to the problem, but these conventions must be followed in order to keep the analysis simple.

The analysis, whether diagrammatic or mathematical, deduces the two laws simultaneously from a simple matching of phases and wavelengths at the boundary. The analysis is continued by applying two of the laws of electricity and magnetism to conditions at the boundary and deducing how much light is reflected from and transmitted through the boundary. The analysis is done for any angle of approach, and the results, sometimes complex, give the intensities of the reflected and transmitted waves. When region two is a mirror, so that no light is expected to be transmitted past the boundary, a special trick can be done with the second index of refraction.

Figure 3-8 simplifies the problem to one of tracking the wave fronts to and from the boundary. The wave fronts are shown simply as a series of parallel lines along the directions of travel, spaced λ/n_1 apart. In region 1 with index n_1, both incident (I) and reflected (R) waves have the same wavelength λ/n_1. In region 2 the wavelength is reduced to λ/n_2. The crux of the analysis is to consider the number of wavelengths that fit into a given advance of position along the boundary and equate the number on each side. An advance s has the component $x = s \sin \theta$ along the boundary, as shown in the inset. The wavelength count is x/λ for this phase-matching procedure.

Along the boundary for the I, R, and T waves, the wave counts are:

$$\frac{s \sin \theta_i}{\lambda_i} = \frac{s \sin \theta_r}{\lambda_r} = \frac{s \sin \theta_t}{\lambda_t}$$

The first equality reduces to $\sin \theta_i = \sin \theta_r$ or $\theta_i = \theta_r$, since $\lambda_i = \lambda_r$. This is the law of reflection. The equality of the incident and transmitted phases is $s/\lambda_i (\sin \theta_i = s/\lambda_t (\sin \theta_t)$. Since $\lambda_i = \lambda/n_1$ and $\lambda_t = \lambda/n_2$, this equality becomes $n_1 \sin \theta_i = n_2 \sin \theta_t$, which is Snell's law of reflection. The laws of reflec-

tion and refraction can also be found with Fermat's principle of least time as shown in the next chapter.

One more result follows from phase matching— all the rays come and go in the same plane, called the *plane of incidence*. This may seem obvious, but it actually requires a proof. The plane of incidence is defined as that plane containing both incident and reflected rays. Since there is no component of the wave moving perpendicular to this plane, the transmitted wave must also travel in this plane. This result can also be seen from consideration of symmetry. Nothing about the incident electromagnetic wave favors the right or left sides of the plane of incidence, so the waves remain in the plane.

When the first medium is less optically dense than the second, the ray or wave is refracted toward the normal as was shown in Fig. 3-8. If the light is incident normally ($\theta_i = 0$), reflection is straight back, and there is no refraction of the T ray. The optical path may just as well be considered in reverse as shown in Fig. 3-9A. A ray approaching in a dense medium is refracted away from the normal. An extreme case is shown in Fig. 3-9B, where the I ray approaches at a certain angle such that the refracted ray, if there is one, travels along the boundary. There can be no greater bending or refraction than this case, and the special incident angle is called the critical angle θ_c. The light energy must go somewhere, however, and it all appears as a reflected ray as shown. For larger angles of incidence, the light continues to reflect completely. The effect is called *total internal reflection* and must occur inside some material that is optically more dense than the material outside the boundary. This is the primary effect enabling use of optical fibers and reflecting prisms as described later.

Experiment 3-1

If you have a block of glass or a prism, look into the block or prism and find the orientation necessary to make an inside surface function as a mirror. The surface will look silvery, and images behind you will be apparent. Then press your finger against the back side of the mirrorlike surface. What happens as you press tighter and tighter? Try to obtain the

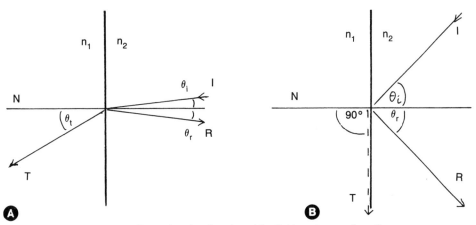

Fig. 3-9. Reversing the direction of the light ray to pass from the medium with higher index of refraction n_2 to the medium with lower index n_2.

effect with any sheet of glass, using a large angle of incidence. ∎

Total internal reflection requires that the index of the material be greater than the index of material outside or behind it. A finger tightly pressed to the surface is optically denser than air and may exceed the internal index. Then the internal reflection is destroyed and reverts to ordinary reflection, which is not very great (about 4% of the intensity). The silvering put on the back (sometimes the front) of a mirror is not making use of critical reflection but rather is an effect of the behavior of metallic substances. They do not have a real index of refraction but an imaginary one. This mathematical concept is rather abstract, even if you have had complex numbers in algebra, but it is a convenient way to describe how a wave ceases waving and damps out in certain materials. Later it will be shown how the top layers of metal atoms can reflect light in an orderly fashion.

3.5 REFLECTED AND TRANSMITTED INTENSITIES

Empirically, each surface of glass reflects about 4% of the light if looked at head on. When incident light shines on a sheet of glass, the results are quite complex, even for the one sheet shown in Fig. 3-10. Reflected light R in the amount of 4% of the original intensity comes from the front surface, and another 4% from the back surface. Any boundary between two optical materials both reflects and transmits light (with the exceptions previously discussed). In this case, air is understood to surround the glass, or the amounts will be different. Ordinary glass also absorbs (A) about 6% of the light, so 86% is transmitted (T') out the back side. At the next level of complexity, the light reflected from the back surface has 4% of its intensity reflected from the inside front surface as R''. It then both reflects from (R''') and transmits through (T''') the back surface. These intensities are now very low, 4% of 4%, or about 0.16% reflection. These secondary and tertiary reflections are not very noticeable, but there is no limit to the effect. A stack of many sheets of glass not touching too tightly will appear as shiny as a mirror because the entering light is returned by a maze of rereflections.

Experiment 3-2

Pick any reasonably smooth surface and look at it at steeper and steeper angles. At some point even a sheet of cardboard will appear shiny when light strikes at a glancing angle on its way to your eye. Also look again at any ordinary sheet of glass (not a silvered mirror) and note the amount of reflection that occurs. ∎

For convenience, Fig. 3-12 showed the light rays at small angles with respect to the normal, and the results will resemble the case of normally incident light. As the angle of incidence grows, results change drastically. Even a sheet of cardboard is a good reflector at steep angles. There must be more to reflection and refraction than simply calculating the angles. There is a strong effect on intensity, and it can be found by considering further boundary conditions as light passes between two materials. To study effects on intensity, the strengths of the electric and magnetic fields must be considered as the boundary is crossed. The boundary conditions are found with all four Maxwell's equations. For transparent optical dielectrics, it is assumed that there is no free charge to move around at the boundary and no currents can flow. However, dielectrics can have net polarization at a boundary—that is, the molecules while neutral might show a separation of charge. The molecules, shown in simplified form in Fig. 3-11, have different shapes and densities on each side of the boundary. An excess of bound charge is the net result. This can affect the electric field normal to the boundary. Usually there is no corresponding magnetic effect.

The results of applying these laws at the boundary are that any electric and magnetic fields parallel to the boundary are unchanged and that any electric fields normal to the boundary are altered according to the net amount of polarization in the dielectrics. Measured constants of the materials— the permittivities, which are related to the squares of the indices of refraction—tell how much. Light waves can pass freely back and forth across the boundary as long as their electric and magnetic components obey these rules. The situation is actually more complex, because it matters which way the electric field is oriented (polarized) as the incident wave approaches the boundary. There are two choices as shown in Fig. 3-12, parallel ($E_{||}$) and normal ($E_{\perp}$) to the plane of incidence (not to be confused with the boundary plane). The plane of incidence contains incident, reflected, and transmitted rays. In neither case are the fields likely to be parallel with the boundary, so components of the fields parallel with and normal to the boundary must be considered. The mathematical details are beyond the scope of this book.

The results are best expressed as reflectance and transmittance coefficients. A reflectance R tells the fraction or percent of incident intensity that is reflected and can be a number between zero and one, or 0% and 100%. A transmittance T tells the fraction of incident intensity that is transmitted

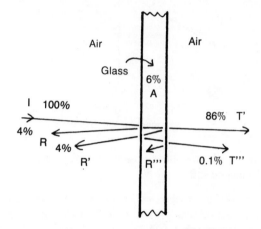

Fig. 3-10. Accounting for all the energy flow through a single sheet of glass, in cross section with absorption A. (The light enters nearly normally, but angles are exaggerated for clarity.)

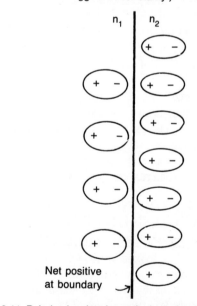

Fig. 3-11. Polarized molecules at the boundary between two media result in an effective net charge.

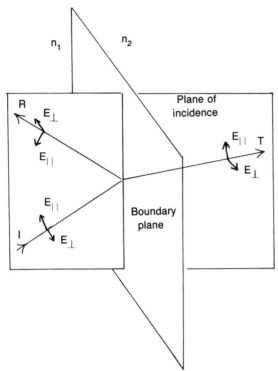

Fig. 3-12. Defining the plane of incidence so that electric field components may be identified as perpendicular to and parallel with it.

through the boundary. These coefficients vary as the angle of incidence varies, and they come in two cases: for incident light with E polarized perpendicular ($E_\perp$) to the plane of incidence and for polarization parallel ($E_{||}$) with the plane of incidence. It also makes a difference which direction the light enters because of the possibility of critical reflection.

By conservation of energy $R + T = 1$; what is not reflected is transmitted (no absorption is assumed). In tracking the flow of energy through reflection and transmission, realize that light striking a surface at an angle has less intensity—less joules landing per unit of area—than light striking normally (headon) as shown in Fig. 3-13. This is taken into account in calculations of reflectance and transmittance. Light transmitted at a smaller angle than it arrived would be more intense for this reason, but other physical reasons may make it less.

The results are best presented as graphs (Fig. 3-14) of reflectance and transmittance as functions of angle of incidence θ_i. Some features should be examined and compared with common sense. These graphs are given for a common case, a boundary between air and glass (index 1.5). Both polarization and direction of travel are crucial and have different

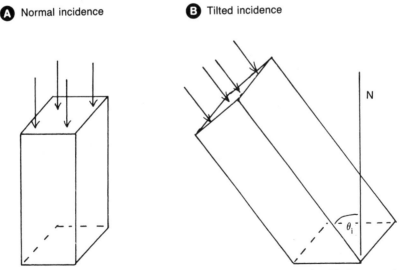

Fig. 3-13. The intensity landing on a surface depends on the angle with the normal.

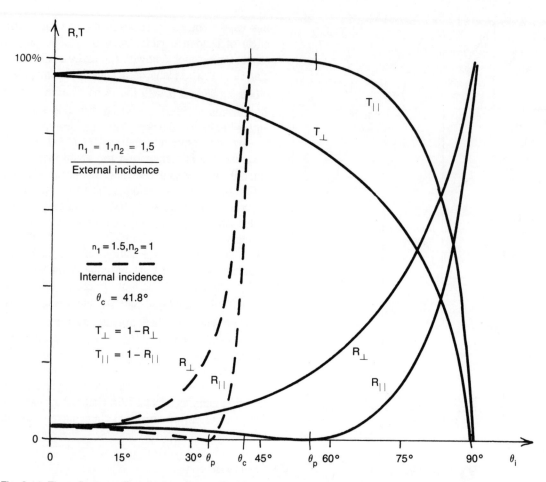

Fig. 3-14. The reflectance R and transmittance T of light at an air-glass boundary, as a function of incident angle θ_i. For clarity, internal transmittances are not shown but are readily calculated from reflectances.

results. For light polarized perpendicular to the plane of incidence, reflectance increases gradually from the low value 4% until steep angles occur. Then it increases rapidly to 100% for light grazing the boundary at $\theta_i = 90°$. Since the total intensity is constant, the transmittance decreases gradually from 96%, dropping more rapidly as the angle approaches 90° and becoming zero at 90°. The somewhat tedious calculations that lead to these results are left to the references (e.g., Hecht & Zajac). For light incident within glass, the indices are reversed and the critical angle is 41.8°. The reflectance and transmittance are shown dashed and climb rapidly to 100% and 0%, respectively, as θ_i increases to 41.8°.

When the incident light has its electric field polarized parallel to the plane of incidence, some details of the results are different. Reflectance $R_{||}$ diminishes to 0% before climbing steeply to 100%. Transmittance $T_{||}$ rises to 100% at a certain angle 56.3°, the same angle at which $R_{||}$ goes to zero. Light incident at this angle is not reflected if polarized in the plane of incidence but rather is completely transmitted through the boundary. The angle 56.3° is called the polarizing angle because incident light, regardless of orientation, is reflected only for the part with field normal to the plane ($E_\perp$). At this angle incident light is converted to polarized light upon reflection. The reason may be seen in a different way in Fig. 3-15. The incident light is refracted

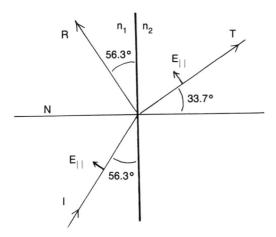

Fig. 3-15. The polarizing angle (Brewster's) is the angle of incidence for which the reflected ray has electric field only perpendicular to the plane of incidence.

to an angle of 33.7°, the complement of 56.3°. $E_{||}$ for the transmitted wave is lined up exactly with the direction of the reflected wave. Therefore the reflected wave will have no $E_{||}$ in this plane. The electrons at the boundary that vibrate to produce the transmitted wave cannot produce any field for the reflected wave. The polarizing angle is also called Brewster's angle because it was found experimentally by David Brewster in 1812. There is a partial polarizing effect for angles near the polarizing angle as the graphs indicate.

Light also undergoes a phase shift of 180° upon reflection. As will be discussed more later, this is a result of action and reaction as incident light encounters electrons at a boundary. One does not notice this looking in a mirror because eyes are sensitive only to the intensity of the reflected light, not its phase (or polarization). Exceptions to 180° phase shift occur when the field is parallel to the plane of incidence and the angle of incidence is less than Brewster's (phase shift is zero), or when the reflection is internal. The latter case is more complex: the phase shift is 180° for $E_{||}$ and zero for $E_\perp$, for small angles of incidence.

Because light is a wave, it does not vanish to zero on the other side of a boundary where there is total reflection. A damped form of the wave leaks past and appears along the boundary as an "evanes-

cent wave." The light wave can be revived if another surface is brought very close to the damped region, within a few wavelengths. Thus light can penetrate otherwise impenetrable barriers. The process is called *frustrated total internal reflection* and has been verified experimentally only in recent decades for light waves.

3.6 SPECTRUM AND DISPERSION

The wavelengths of light are rather small numbers, and the frequencies are rather large. Red light has the longest wavelength in the visible, about 700 nanometers (billionths of a meter, abbreviated nm). Longer wavelengths are infrared, measured in 1000s of nanometers (micrometers). Orange light is about 600 nanometers, yellow about 580 nanometers, green about 550 nanometers, blue about 450 nanometers, and violet about 400 nanometers. In between colors have in between wavelengths. Optics is rarely concerned with ultraviolet shorter than about 100 nanometers. From the relation $c = \lambda f$ the equivalent frequencies can be calculated, ranging from about $4(10)^{14}$ or 400 terahertz (THz) for red to about $7(10)^{14}$ or 700 THz for violet.

Each source of light emits a distribution of colors, usually with a peak at some wavelength. The sun emits about 50% infrared, about 45% visible light, and about 5% ultraviolet light, in terms of energy. The wavelengths at which the most energy is emitted correspond to green at about 550 nm. The sun does not appear green because it emits large amounts of all other colors, and the resulting mixture is a yellowish white. Other stars might be reddish or bluish. Hotter stars emit more light at shorter blue-violet wavelengths, and cooler stars emit more reddish light. It is probably no accident that the eye is most sensitive to green light at about 550 nm, the same as the sun favors.

Dispersion is the dependence of the index of refraction—and therefore the speed of light—on the color or wavelength that is being used. Generally the shorter wavelengths (blue and violet) travel slightly slower than red light in materials. The effect is traceable to the way that the vibrating electric field in light makes the electrons vibrate in

Fig. 3-16. An electron attached to an atom as if by a spring generates an electromagnetic wave when it vibrates.

materials. The electrons in atoms might be modeled in a crude way as connected by tiny springs to the atoms, so that they have their own natural frequencies of vibration (Fig. 3-16). In glass the natural vibrations are faster than visible light. In the ultraviolet range, the light vibrations are opposed by the electrons in the glass, and the light is absorbed. The electrons are shaken faster than they want to be. Glass appears more and more opaque as the wavelength goes farther into the ultraviolet.

Materials whose natural electron frequencies are close to the frequencies of light impinging upon them will strongly absorb the light. Light below the natural frequency is reradiated when the electrons are shaken because the electrons can radiate in phase with the light. Light thus travels through transparent glass. The closer the light frequency is to the natural frequencies of the material, the more it interacts. The result is lower light speed, expressed as a larger index of refraction. In the infrared range, the atoms and molecules as a whole are vibrated slowly enough to respond to light, providing other, lower natural frequencies. The oxides in glass absorb infrared so that glass is nearly opaque to infrared light. Chapter 9 gives more description of these interactions of light and atoms.

3.7 MORE OF THE PHYSICS AND MATHEMATICS OF LIGHT WAVES

More mathematical results and physical principles for the Maxwell electromagnetic theory are given here for ambitious readers. Maxwell's equations have several forms, always involving calculus notation. Therefore, they are left to the references. The wave equation that results when Maxwell's equations are manipulated must also be stated in terms of calculus. However, the solution is relatively simple and has two parts, electric and magnetic. It is given here for a plane wave traveling along the positive z-axis:

$$E(z,t) = E_0 \sin (2\pi z/\lambda - 2\pi ft)$$

$$B(z,t) = B_0 \sin (2\pi z/\lambda - 2\pi ft)$$

Vector notation has been omitted. The amplitudes E_0 (in newtons/coulomb) and B_0 (in webers per square meter) are not independent but related as $E_0 = cB_0$. The time-averaged energy density u (in joules per cubic meter) of the wave is:

$$u = \epsilon E^2/2 + B^2/2\mu$$
$$= \epsilon E^2 = B^2/\mu$$

The latter relations hold because c can be calculated from permittivity ϵ and permeability μ by:

$$c = 1/\sqrt{\epsilon\mu}$$

In space the vacuum values ϵ_0 and μ_0 must be used. In most optical situations, μ has the vacuum value.

The power flow or energy flux S in joules per square meter per second is called the magnitude of the Poynting vector (after Henry Poynting) and flows in the direction the wave travels. It is given by:

$$S = EB/\mu$$

If the total energy U of a light wave is known, then the momentum p it carries can be found from:

$$p = U/c$$

The critical angle θ_c for total internal reflection can be found from Snell's law by using n_1 greater than n_2 and supposing θ_t is 90° so that $\sin \theta_t = 1$. Then $\sin \theta_c = n_2/n_1$.

The polarizing (Brewster) angle θ_p can be found by applying $\theta_p + \theta_t = 90°$ to Snell's law to obtain:

$$\tan \theta_p = n_2/n_1$$

For light incident normally to a surface, the reflectance reduces to a simple expression and is the same for light of any polarization:

$$R = R_\perp + R_{||} = (n_2 - n_1)^2/(n_2 + n_1)^2$$

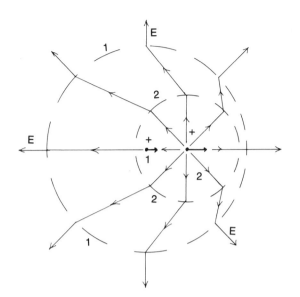

Fig. 3-17. Accelerating a positive charge from one position and speed to another produces a kink in its electric field that propagates outward.

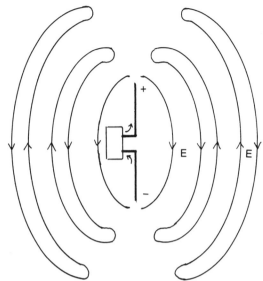

Fig. 3-18. Pushing charge back and forth with a generator in a dipole antenna causes loops of electric field (and magnetic, not shown) to break loose and propagate outward as waves.

The electric field that radiates away as a result of shaking an electron of charge e with a sinusoidal force at frequency f so that its acceleration is $a_0 \cos 2\pi ft$ is:

$$E(r,\theta) = (-e \sin \theta)\,(a_0 \cos 2\pi f(t - r/c))/4\epsilon_0 rc^2$$

The field is symmetrical around the direction of electron motion, and θ is measured from the axis along which the electron oscillates. The quantity $t - r/c$ is the retarded time that applies at the location of the electron. The resulting frequency of the wave is the same as the frequency with which the electron is vibrated.

As shown in Fig. 3-17, the acceleration of an electron produces a transverse kink in the electric field around it. In position 1 there is a set of radial electric field lines pointing to where the electron was. In position 2 there is a new set closer and displaced. Without worrying about details, there must be transverse kinks connecting these sets of field lines to make them continuous. Moreover, whereas the strength of the radial electric field decreases with the inverse square of the distance, the strength of the field in the kinks decreases inversely with dis-

tance. This raises the possibility of sending as much energy as desired an infinite distance, as a wave can do. Similar considerations apply to the generation of the magnetic component of the wave.

Another common way to generate an electromagnetic wave is with an oscillating *dipole*, an arrangement of a positive and negative charge whereby the distance between them varies sinusoidal. This can be done by pushing charge into one part of an antenna while pulling it out of the other side (Fig. 3-18). When the excess charge is then pulled back to the other side, the electric field (and magnetic field) connecting the two sets of charge is pinched off to form closed loops that travel outward through space. A self-propagating electromagnetic wave is born. The spacing between the loops is the wavelength. The wave has different forms in two regions, the near and the far. Near the source, electric field seems to point outward. Far away it vibrates transversely. The power radiated by an oscillating dipole is proportional to the square of the frequency. Thus high frequency oscillators can send electromagnetic waves out much more efficiently than low frequency ones.

33

4

Geometric Optics

Two principal and related applications of optics are forming images and transmitting information. Geometric optics, where light rays are traced through an optical system, is one model for optical design. Another related model, wave optics, provides perhaps a better explanation of the physics but is more complex to use for design. Wave-based applications are deferred to later chapters where ray methods prove inadequate. The basis of geometric or ray optics is that light travels in straight lines (rectilinearly, except in unusual media). Principles developed for geometric optics seem to explain the relevant phenomena adequately, and the mathematical methods are particularly simple, often requiring only geometry. But keep in mind that a more exact model of light is an electromagnetic wave that can give these same simple results in quite complex ways, as introduced in the preceding chapter.

4.1 ILLUMINATION AND SHADOW

One might think that objects are either illuminated or dark—in shadow. But observation around us shows that nothing is illuminated so brightly or so uniformly as to show its features regardless of variation in illumination. No object in shadow is pitch black. Because we live in air, which is a source of diffuse light, and are usually surrounded by objects with varying reflective capabilities, no shadow is fully dark. Where one light source is blocked, others fill in.

For one extended light source (not a point), the resulting shadows consist of two parts, the *penumbra*, where illumination comes from part of the source, and the *umbra*, which is darker and receives no direct illumination (Fig. 4-1). In an eclipse of the sun, a few observers find themselves in that part of the shadow where the moon appears to completely cover the sun. Many more observers are located where they can see part of the sun. An aerial view of the ground would show a small dark circle or ellipse a hundred or so kilometers across, surrounded by a larger, less dark penumbra.

Were the moon not quite as close to the Earth, its shadow would not reach the Earth. Similarly for all light sources that are bigger than the barriers in

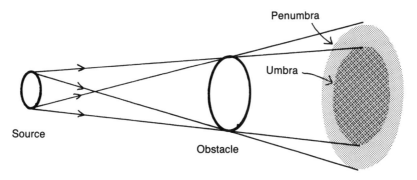

Fig. 4-1. Structure of a shadow when the light source has extent.

front of them—the sun being the largest of all—there is some distance behind the barrier beyond which the umbra is gone. As the distance increases, the penumbra becomes weaker, too. A bird passing in front of the sun would hardly be noticed (if the bird were flying very high above the observer). On the other hand, a very small bright source would produce a very long umbra.

Experiment 4-1

On the ground under a well-leafed tree, the sun comes through small breaks in the leaves in rare places. Put a smooth sheet on the ground to show that the spots of sunlight are not irregular like the holes among leaves but rather are all round, like images of the sun. ■

4.2 FERMAT'S PRINCIPLE FOR LIGHT RAYS

In any optical situation, of all paths that light might take from one point to another, light will take the path that requires the least time. This is Fermat's principle, although it was first thought of in ancient Greece. This principle can be used to obtain the laws of reflection and refraction in a different and simple way. Lest it be thought that there are many independent routes to Rome or Greece in physics, there are fundamental connections among the several methods for obtaining these laws. For Fermat's principle, it is sufficient to model light with rays.

In Fig. 4-2 light rays are conjectured to take several different paths from source S to observer

O in the process of reflecting from a mirror. Because the light spends all its time in the same medium, the shortest distance also gives the shortest time. It should be noted that the observer sees not S but S', an image source located as shown. The bent paths from S' to O are clearly not the shortest paths, but the straight path is shortest in this case. It happens to be the path with $\theta_i = \theta_r$, proving the law of reflection.

In Fig. 4-3, a light ray is shown taking the path SA in the first medium, which may be assumed to have a low index of refraction, and then path AO in a medium with a higher index. The bent path already shown must be close to correct because the light travels more distance in the higher-speed medium while using less time, then enters the lower-speed

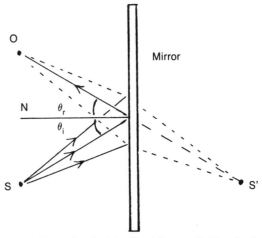

Fig. 4-2. Fermat's principle of least time applied to reflection.

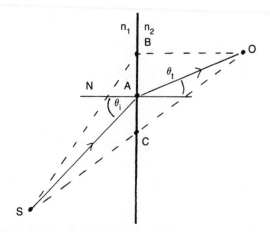

Fig. 4-3. Fermat's principle of least time applied to refraction.

medium at an advantageous point A that results in less time traveling a smaller distance. Point B is not a better entry point because the time spent in the first medium is then quite a bit longer with little gain from the slightly shorter path in the second medium. Point C, with no bending, is not correct because the distance in the second medium is much larger.

The distance for the path SAO can be calculated from the geometry. The shortest distance can be shown with elementary calculus to be that which makes $n_1 \sin \theta_i = n_2 \sin \theta_t$, which is Snell's law. Looked at in this way it may seem strange that light knows which way to go in the first medium to strike the boundary in just such a place as to end up at O. But a deficiency of the ray model is revealed. The better model, waves, has light striking all along the boundary at the same angle θ_i. Inside the second medium the light travels at the new angle θ_t, encountering point O as well as many other points.

Fermat's principle has many other practical applications. Warm air is less dense, so the speed of light is a little greater in it. This leads to atmospheric effects such as seeing the sky as a mirage on a hot road. Light from the sky can take less time to reach an observer by bending to pass through the less dense air. The thinning of the Earth's atmosphere causes the setting sun to be seen longer than it is above the horizon. With Fermat's principle, the ideal mirror shape for collecting starlight and bringing it to a focus can be predicted to be a parabola. The ideal shape for a lens that can gather many light rays

and bring them to a focus can be predicted. These phenomena are explained by other means later, but least time is one way to explain them.

4.3 THE PINHOLE CAMERA—OBJECT AND IMAGE

One goal of practical optics is forming an image of an illuminated object. The image may be seen directly by the eye, having formed on the retina. Or it can be formed on a smooth, white screen, where it appears as a glowing miniature of a real scene. Or it might be formed on photographic film designed to capture a permanent record of the image. It is usually desired that the image be a sharp, clear replica over all parts of the image, and that the colors all be accurate and focused. This is a difficult achievement.

The simplest way of forming an image of an object is to use the pinhole camera method. The method was known to Aristotle in ancient Greece and described in the Renaissance by Leonardo da Vinci. It was called the *camera obscura*, which is Latin for "dark chamber." The device is simple and is the forerunner of both photographic cameras and the use of lenses to form images.

An early reason for trying to form an image was to trace or otherwise duplicate a real object more accurately in art, as described by Giovanni della Porta in the 16th century. Early artists found that light rays from a brightly illuminated object would, when passed through a small hole and permitted to fall on a flat surface in a dark room or box, form a small, inverted, and surprisingly sharp image of the object (see Fig. 4-4). For illustrating geometric optics, an object in the shape of an arrow is often used so that any inversion of the image will be readily apparent. The image will be dim unless the object is in full sunlight. And the smaller the pinhole, the less total light can pass, so the image is still dimmer. Unfortunately, making the hole larger for more light results in a less sharp image.

Light rays from a point A on the object travel in all directions, and a few go in those directions that can pass through the hole. These land on the screen in a variety of places. Point A is smeared out to region A' on the screen. This same process occurs

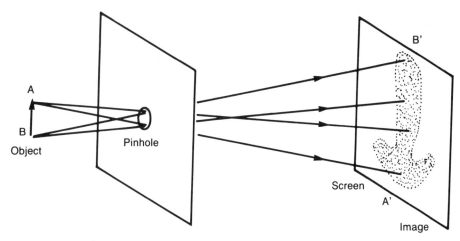

Fig. 4-4. Unless very small, a pinhole will not produce a sharp image of a bright object. (Hole and image exaggerated here.)

for all illuminated points of the object. Rays from point B, at the other end of the object, that pass through the hole are similarly spread over a region B'. They arrive at a place inverted from the location of point B on the object. Thus a pinhole imaging device *inverts* the image. Up becomes down and left becomes right. If the hole is enlarged sufficiently compared to the distance of the object, the rays from any point of the object spread so much as to cover most of the whole image. The image becomes too blurred to be useful.

Experiment 4-2

Punch a very small pinhole in cardboard and use it to image the sun on another cardboard on the ground. Use a large cardboard for the hole so that there is a shadow in which to observe the image. Do not look at the sun. What happens as the barrier cardboard with the pinhole is moved farther from the screen cardboard on the ground? Gradually enlarge the hole until the image of the sun is blurred. How big must the hole be before the sun is just a blur? What happens if the hole is much bigger—no barrier is used at all? Perhaps you should wonder why objects do not form images on surrounding surfaces without any help at all. ■

The sun is so far away that there is a tendency to assume it sends parallel rays of light. If rays are

parallel, then a pinhole only forms an image that is a point. However, the pinhole does form an image of the sun. The sun is not a point in the sky but an object of substantial size covering about one-half degree of arc. Therefore, rays from various points on the sun can take different paths and form an image. This is a safe way to observe a solar eclipse and has been so used for at least a millenium. The image can be made larger by moving the screen away, but then it becomes dimmer. As the pinhole is made larger, its size eventually has no effect, as if there were no barrier. We know the sun shining on cardboard on the ground does not form an image of itself but uniformly illuminates it like everything else.

Experiment 4-3

Start with a fresh pinhole to form images of the sun, but this time make several unusual shapes, triangular, square, or even a cutout of one of the letters of the alphabet. Work small and keep the hole size less than 1 centimeter. Do you obtain a bright geometric shape on the screen, or something else? Be sure to test these holes far from the screen. ■

If the holes are close to the image, the sun shining through holes of various shapes gives the equivalent of shadows in reverse—the shapes are bright and resemble the holes. But if the holes are moved sufficiently far away, each hole and each part of a hole seems to give a round, blurred image. Go sufficiently far, and a hole the shape of the letter E will give a round image of the sun, although sharpness will not be as good as that for a round hole. A task later in this book is to determine from basic principles just how far away the hole must be to function as a pinhole regardless of its shape.

The pinhole camera should not be thought an historic gadget because it is in use today by serious and scientific photographers. In some conditions it produces the sharpest possible image with negligible distortion and devoid of any color errors. Its depth of field is enormous, which means that all parts of a scene are imaged sharply regardless of how close or how far they are. Only the eye can perform better. If the hole is made too small, the wave nature of light interferes and prevents sharper imaging. Curious photographers should be forewarned that the effective f-number is about 500, indicating extremely weak exposures. The optimum pinhole size has been found to be about 0.5 mm for images about 25 cm from the pinhole.

4.4 DESIGNING A THIN LENS

A common lens could be chosen and studied to determine how it works according to the laws discussed thus far. Or the laws can be applied in a more basic way to deduce what shape a lens must have in order to do what is wanted of it. The purpose of a lens is to gather light from an object and *focus* it to form a sharp image. Rays that initially diverge in various directions from one bright point on an object are said to be focused when the lens causes them to pass through another point (the *conjugate* point) somewhere beyond the lens.

In deducing the required form of a lens, a blackbox approach will be used. In Fig. 4-5, the lens is shown as a blackbox or unknown device placed between the object and the screen. The goal is to invent a device to put in place of the blackbox that will take light from a point A on the object and form a good bright image as the point A' on the screen. Whether the image is inverted or not is not of concern for now. This analysis is a form of a very general approach called input-output analysis. A transform or transfer function is sought that will convert the object information into image information.

To improve upon the pinhole, more light rays must be gathered from the object. Therefore, the opening or aperture must be large. Hence the blackbox or transform has been drawn as an open cylinder, like a food can with both ends removed. What must be put in the cylinder to focus rays? The principle of least time proves valuable here. A device is needed that provides the same time of travel for

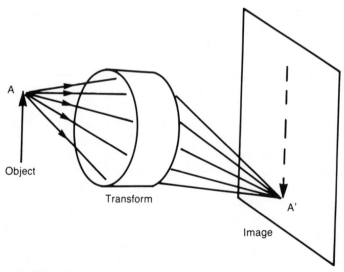

Object

Transform

Image

A'

Fig. 4-5. Designing a lens as a transform that converts object to sharp image.

many rays from point A to point A'. Different paths have different lengths, as shown. A use of refraction is indicated. A clear material with a lower speed of light could introduce the proper delays so that all paths take the same time. Another requirement is that outlying divergent light rays must also be bent by the device, a consequence of refraction at a boundary between air and an optical material. The goal is not to cause all light waves to arrive exactly in step. Rather it is to provide a device that makes all light paths effectively the same for rays from A that are to focus at A'.

A sheet of clear material such as glass which is thin at the edges and thick at the center might satisfy the requirements. Rays passing through the center would ordinarily have the shortest path. They need the most delay by passing through the thickest glass. Rays near the edges need less delay in order to keep up, but some must be bent sharply to arrive at the proper image point. By refining this approach, an exact calculation of the necessary contour of the glass device can be made, giving the form shown in cross section in Fig. 4-6A. The device is a *lens*. Another way to see what is needed is to imagine a stack of little glass blocks of various shapes as shown in Fig. 4-6B. The thinnest blocks at top and bottom cause the least delay of light. The thickest are at the center and give the most delay. A necessary refinement of the glass blocks is to slant their sides, forming prisms that give the necessary bending, too.

Lenses are most easily manufactured with spherical surfaces. This does not mean that they are round, although many are as another manufacturing convenience, but that their surfaces resemble the surfaces of spheres of suitable radii. Another approach to lens design is to study where rays are bent when passing through a spherical boundary between air and glass or glass and air. Spherical shapes are easy to analyze and produce a simple answer in the case of *paraxial* rays almost parallel to the axis through the center of the lens and close to it (Fig. 4-7A). The regions far from the axis are best avoided. The spherical shape is an approximation of a family of special shapes that satisfy the requirements of focusing certain kinds of rays. When analyzing an air-glass boundary, note that there is nothing wrong in principle with having a focal point inside the glass. In most situations, rays will be focused to other points outside the glass.

4.5 LENS LAWS FOR THIN LENSES

The result of analyzing where rays focus after passing through the spherical boundary shown in Fig. 4-7B is:

$$n_1/x_o + n_2/x_i = (n_2 - n_1)/R$$

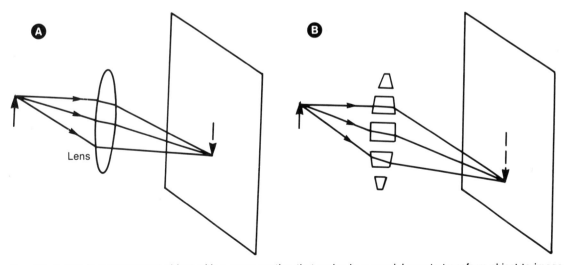

Fig. 4-6. A lens is a transparent object with a cross section that varies in a special way to transform object to image.

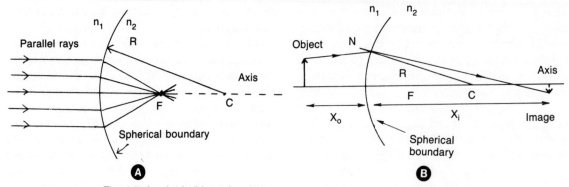

Fig. 4-7. A spherical boundary between air and glass (shown in cross section).

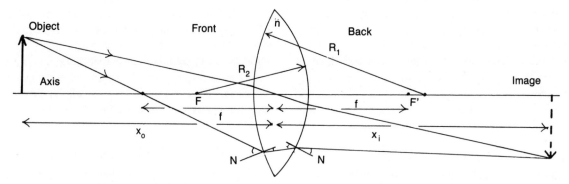

Fig. 4-8. A thin lens is made of two spherical surfaces (exaggerated for clarity).

Here R is the radius of the spherical surface, x_o is the distance from object to surface as shown, x_i is the distance from surface to image, and n_1 and n_2 are the indices of refraction of air and glass, respectively. Again, the assumption is that rays are paraxial. To the extent that this approximation is not followed, the resulting image will be poorly focused. This relation tells where the image is when the distance to the object is known, and vice versa. All distances are measured along the axis.

What happens to parallel rays is always of interest. If the object is moved to infinity so that x_o becomes infinity, rays from it will be parallel with the axis. The image is then formed at a special point called the focal point F, and its location a distance f from the boundary can be calculated to be:

$$f = n_2 R / (n_2 - n_1)$$

Any two spherical surfaces can be combined to form a lens with index of refraction n as shown in Fig. 4-8. This shape of lens is called *convex*, and its action is *converging* or positive. The location of the image from the second surface can be found by assuming that the image from the first surface is the object for the second surface. The general relation works regardless of whether rays are converging to form a *real* image or diverging to appear to come from a *virtual* image. Other possible lens shapes as shown in Fig. 4-9 are possible so long as the lens grows thicker at the center in the proper way. The lens orientation is irrelevant in noncritical applications.

The result of combining two imaging calculations is to locate the final image with this relation:

$$1/x_o + 1/x_i = 1/f$$

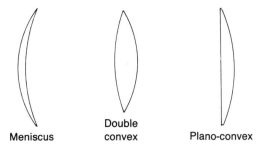

Fig. 4-9. Possible shapes for convex converging lenses.

where x_o and x_i are measured from the center of a *thin* lens.

The focal length f can be calculated from the two surface radii by:

$$1/f = (n - 1)(1/R_1 - 1/R_2)$$

The glass has index n and air is assumed outside. Radius R_1 is positive if the center of curvature is located opposite entering rays, as shown in Fig. 4-8. R_2 is negative because the center is on the front side of the lens. If the radii did not have different signs, a symmetrical lens would have no converging power. Thus a lens with two boundaries can be modeled as an abstract object with focal points F and F' on either side, each a distance f from the lens regardless of the difference in curvature on its two sides. The two relations constitute the Gaussian lens formula. The lens has been assumed thin so that its thickness is negligible compared to the object, image, and other distances. The lens has the same effect regardless of which way light rays pass through it. There are practical limits on how short the focal length can be. When the focal points are close, the radii are small, and the lens must necessarily be small. An extreme case is a clear glass marble. Whatever size it has, the focal points are the same distance outside its surface as its radius.

The detailed behavior of any light ray passing through any lens follows the law of refraction as was shown in Fig. 4-8 for a ray. At the first surface the ray is bent toward the normal. Then it approaches, from inside the lens, the second surface at some new angle. It emerges refracted away from the normal. Generally, if the lens is a converging one, the over-

all effect on the ray is to bend it toward the axis. The simplified model for a lens supposes there is one plane that replaces the lens and at which all bending occurs. Objects sufficiently far away can be much larger than the lens, and the optical performance will still be adequate.

At the center of the lens, on the axis, the lens is effectively a sheet of flat glass. The ray is bent, then bent back the same amount. It passes straight through, unchanged in direction but with a small lateral displacement. A flat sheet of glass is a lens with zero curvature and therefore with focal points infinitely far away. A lens with one flat surface can still have two focal points reasonably close together. The two focal points are not each assigned to one of the surfaces but rather result from the lens as a whole.

Another form of the lens formula is convenient to use. If the distance from object to F is defined as x, and the distance from F' to image is defined as x', then object and image distances have this relationship:

$$xx' = f^2$$

This Newtonian relation can be calculated directly from the Gaussian one. It is easily proven geometrically as shown in Fig. 4-10. This diagram also illustrates the geometric method of locating images by ray tracing, discovered by Robert Smith about 1738. For purposes of ray tracing, the object is assumed to be a source of light with every point of it emitting rays in all directions. The rays traced through the lens system need not be rays that a real object can actually send through the lens. The thin lens is shown simply as a vertical line representing a cross section of the lens plane where all refraction occurs. Analyzing lens behavior in terms of rays should never be done by guesswork. There must always be a reason for a particular ray, based on the principles given here.

In Fig. 4-10, the ray from object point A to image point A' that passes through the center C of the lens is refracted the same on both sides and is straight and essentially unchanged. A ray from the object and traveling parallel with the axis must be

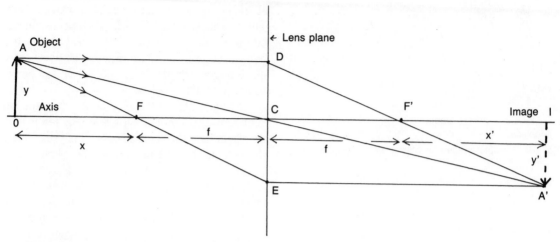

Fig. 4-10. Finding the location and size of image by ray tracing and geometry for a thin converging lens.

bent by the lens at D in such a way as to pass through the focal point F' on the other side. Two separate rays traced are sufficient to locate the corresponding image point A' where they intersect. One more ray is easy to trace, the one from object through first focal point F. As it emerges from the lens at E, it is bent to run parallel to the axis. It also passes through A', as would all other rays from A. The triangles AOF and FCE are similar so that the ratios y/x and y'/f are equal. The triangles DCF' and F'IA' are similar so that the ratios y'/x' and y/f are equal. Combining these to eliminate y and y' gives the Newtonian formula.

The effect of a lens on object size can also be calculated. The *lateral magnification M* is defined as the ratio of the image height y' to the object height y. The image height, measured from the axis, will be larger or smaller than the object height by an amount M, where M can be larger or smaller than one. From the preceding geometry or from similar calculations expressions for the magnification can be found. It is important to keep track of which side of the axis the image is on. If y is taken as positive, then defining $M = y'/y$ provides a negative value for M when the image is inverted to the other side of the axis and y' is negative. The results for the magnification have several forms:

$$M = y'/y = -x'/f = -f/x$$

The last form is the most useful.

The various lens relations give results that are easily interpreted if attention is also paid to algebraic signs. A real image is a positive distance beyond the second focal point F'. The image is virtual if either the lens is a *diverging* one (Fig. 4-11A) or if the object is located closer than the focal point (Fig. 4-11B). Then a negative result for x' is obtained and must be interpreted as measured the other way from F'. The magnification for a virtual image is positive, showing it is noninverted. An object closer than F to a converging lens is assigned a negative distance x measured inward from F. The virtual image is drawn dashed to indicate its ghostly existence and is usually uninverted (erect). It can be found by ray tracing by following the preceding rules strictly. A converging lens lets rays diverge if the object is so close to the lens that its limited refracting power cannot bring them to a focus, even far away.

A diverging lens is assumed to have f negative and is called a *concave* or negative lens. One or both radii are negative for a concave lens such as was shown in Fig. 4-11A. The radius for the front surface has its center at the front and similarly for the back radius, so that negative values must be used in calculation. Object positions are negative if measured behind the front F, as are image positions measured in front of the back F'; this is the same convention as for converging lenses. A diverging

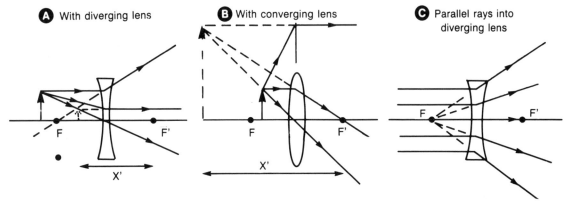

Fig. 4-11. Ways to obtain a virtual image and diverging rays.

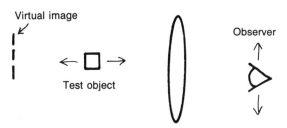

Fig. 4-12. Using parallax to find the distance of a virtual image (top view).

lens causes parallel rays to appear as if they came from the focus on the other side (front) of the lens (Fig. 4-11C).

A virtual image cannot be seen unless the converging power of the eye is strong enough to bring the diverging rays to a focus on the retina. It will not form on a screen. The rays for a virtual image have order that defines an image. But, the image can be seen only by letting the diverging rays fool the eye into thinking that they came from an image when in fact they are bent outward from the object rays.

Since a virtual image cannot be located on a screen, the question arises of how one can measure its distance. One method uses *parallax*. This is the apparent shift of position of an object in the foreground as compared to the background when one's head moves from side to side. When looking through a lens and perceiving a virtual image, it will be at the same distance as a test object (Fig. 4-12) when there is no parallax effect. The test object is

moved back while the observer moves from side to side. At some distance, moving the head will show no relative shift of image and test object, and this is the image distance.

Experiment 4-4

If you have a lens or magnifying glass, the focal length can be measured by choosing a distant light source—streetlights a kilometer away or a sunlit scene—and forming an image on cardboard in a darkened room. Now, test by placing your eye near the lens but inside and outside the focal point. Move it back and forth and examine the image you see for its virtualness. ∎

Experiment 4-5

A good object for close work is the filament of a light bulb. Find out where lens and screen must be placed to form an image the same size as the object. Is this also the situation for having object and image as close together as possible? Also diagram this situation, and use ray tracing (and calculation) to find or check the results. ∎

At first glance, the use of a lens to obtain a magnification of 1 (really −1) seems useless. But suppose lifesize (true scale) photographs of objects are wanted. By symmetry, when object and image are the same size, they are each the same distance from the lens. In this case, each is a distance $2f$ from the lens, and object and image cannot approach each

43

other more closely. When one is brought closer than $2f$, the other recedes.

The importance of the indices of refraction for lenses is illustrated by considering a lens under water made with a thin plastic box filled with air. If the box is in the shape of a convex lens, the actual behavior of the lens will be diverging. Now the dominant index of refraction (about 1.33 for water) is outside the lens. Similarly, a box of air in the shape of a concave lens will act converging.

Numerical examples using the lens relations, as well as further examples of ray tracing, are given in section 4.8 where lenses are combined.

4.6 EXTENDED OBJECTS

It is not convenient to do optics using only paraxial rays to avoid inaccuracies. The price paid

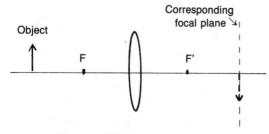

Fig. 4-13. The focal plane for an object ideally is flat, and its location depends on the object distance.

for using larger objects and more of the lens is noticeable distortions in the images. The next chapter shows methods of curing the various distortions that arise. Even if the whole lens is used, the brightness that can be achieved is limited by the area of the lens.

Usually it is desired that rays from all parts of a nearby object fall in one plane, the *focal plane*, located beyond F' as shown in Fig. 4-13. It contains the focal point of the lens only for objects at infinity (astronomical objects like stars). Having it planar is especially convenient if the image is to be captured on film, because spherically curved film is nearly impossible to deal with. Having all focusing occur in a plane is also desirable for applications such as projecting movies on screens. To a first approximation, the rays from a flat object tend to be focused on the flat focal plane.

Real objects are usually three-dimensional. Rays can originate from points at various distances from a lens. The lens, with some accuracy, can bring these rays to focus in three-dimensional space—that is, to a series of focal planes. Often one does not want a three-dimensional image but a two-dimensional one (Fig. 4-14). Rays from various parts of a tree, for example, should all focus in one plane to form a flat image of the tree. No lens can do this perfectly. There is a depth of focus within which a

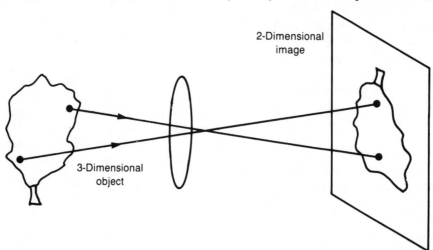

Fig. 4-14. Imaging three-dimensional objects on a two-dimensional screen requires depth of focus which a fixed lens cannot provide.

sharp image of various parts of the object is obtained. More distant parts of the object will be imaged poorly onto one plane.

To trace what a lens inverts and does not invert when an object is transformed to a real image, consider the three dimensions represented by three fingers of the right hand. Let the thumb and second finger point at right angles in the object plane. The image will have each of these reversed. The third finger should point along the optical axis at the lens. Then the image will have that finger pointing away from the lens (direction unchanged). Reversing one finger would create a left hand. Because the lens reverses two fingers, the image is right-handed like the object.

Experiment 4-6

Try the preceding work with the right hand in your mind or on paper unless you can set up a lens to try it. ■

The locations of object points toward or away from the lens are also subject to magnification. This longitudinal magnification works out to be negative and equal to the square of the transverse magnification.

4.7 APERTURES, STOPS, AND BRIGHTNESS

An *aperture* is nothing more than an opening through which light can pass. Apertures are important in optical systems to block unwanted light and restrict images. In a complementary way, an aperture is a *stop*. Light beyond the aperture is stopped from entering. The aperture automatically controls the amount of light entering the system and therefore the brightness of the image. The simplest use of an aperture is to limit light rays to the middle of a spherical lens, thus improving the quality of images. When close to the lens, it is called an *aperture stop* and is shown in cross section as AS in Fig. 4-15. It functions as an *entrance pupil* because the ray shown is the last one which will pass the stop. Looking into the lens, the maximum effective aperture is that indicated as the entrance pupil; it is a virtual image of the AS.

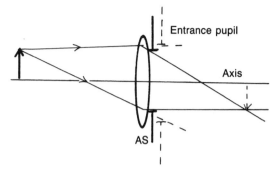

Fig. 4-15. Aperture stops (AS) can be put in several places to control stray light and the amount of lens used. Here AS is also the exit pupil.

More important perhaps is the *exit pupil*, which determines the largest possible image for the system. Looking backward into the lens, one sees in this case that the aperture stop limits the image size. If the aperture stop were placed in front of the lens, then roles would be reversed. It would serve as the entrance pupil, while the exit pupil would be determined by the direction of the outermost rays that can pass it. Then the exit pupil is a virtual image of the AS. The AS can be called an object just like any other, and its images could be found or computed accordingly.

Variable apertures are called *diaphragms*, usually made of thin, blackened metal leaves in an interleaved pattern. Moving a lever causes more or less overlap of the leaves, varying a polygonal aperture in the middle. Another vital method of stopping unwanted light in an optical system is to paint all interior parts flat black. The depth of field depends strongly on the aperture, just as the size of a pinhole determines how sharp an image is. A small aperture prevents a wide range of ray path lengths from various parts of the object to the image plane.

Images tend to be dimmer at their edges because the finite sizes of the lens(es) and any AS prevent as full a set of rays from the outer parts of the object to be processed by the lens. The brightness of the image is proportional to the area of the lens and therefore proportional to the square of the lens diameter D. The major portion of the light in an image passes through the outer parts of the lens where most of the area is.

The size of the image, over which the light is spread, is proportional to the square of the magnification M, which depends on f. Therefore, the brightness is proportional to $(D/f)^2$. The reciprocal f/D has been defined as the *f-number* of the lens—the ratio of the lens focal length to its diameter. A small f-number indicates a lens that can gather much light, just as the most expensive camera lenses have low f-numbers like 1.4 or even 1.2. This subject will be developed more in the chapter on applications.

4.8 COMBINING LENSES— WITH AN EXAMPLE

Two thin lenses may be combined for the purpose of forming a more useful instrument as will be shown in the chapter on applications. An example is a simple telescope. The optical procedure is to use the image formed by the first lens as the object for the second lens. The results can be found by experiment, by ray tracing, or by calculation.

In Fig. 4-16, a small, distant object is placed 60 centimeters from F_1 for a thin converging lens 1 with $f_1 = 30$ centimeters. The image is calculated to be at $x'_1 = f_2/x_1 = 900/60 = 15$ centimeters. Since x'_1 is positive, the image is real and located outside F'_1. Diverging lens 2 with $f_2 = -30$ centimeters is placed nearby as shown so that the first image is 20 centimeters behind its F_2. Then $x_2 = -20$ and $x'_2 = 900/-20 = -45$ centimeters. The image is located 45 centimeters in front of the back focal point F'_2. The two rays needed to locate the first and second images are also shown, along with

a third ray easy to construct. Note that the final image is virtual (seen with diverging rays), and that the ray method provides its size as well.

The magnification of the images can also be calculated. The first image is magnified by $M = -f/x = -30/60 = -1/2$. The height of the first image is the object height of 20 centimeters multiplied by $-1/2$ to give -10 centimeters. The magnification to be applied to this image as object to obtain the second image is $M = -(30)/(-20) = +1.5$. Thus the final image is $(-10)(1.5) = -15$ centimeters tall. The second virtual image thus has the same orientation as the first image, as expected.

When combining lenses that are close together, the total effective focal length can be calculated by adding the *powers* of the lenses. The power of a lens is simply the reciprocal of the focal length measured in meters. The units are diopters. A diverging lens has negative power. Because the surfaces of one lens are close together, a power for each surface can be calculated and the powers of both surfaces added to obtain the power of the lens.

When combining two lenses, 1 and 2, separated by a distance d, the total effective focal length of the combination can be calculated by:

$$1/f = 1/f_1 + 1/f_2 - d/f_1 f_2$$

When a second lens is placed at the focal point of a first lens (Fig. 4-17), there is no effect on the combined power. All rays either pass through the center of the second lens where they are unaffected,

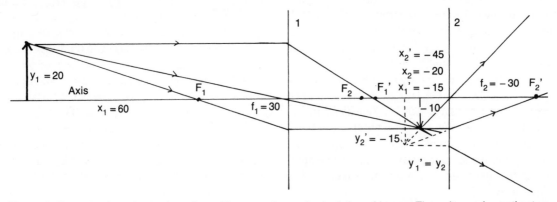

Fig. 4-16. Example of two lenses in series, with ray tracing and calculation of image. The units are in centimeters.

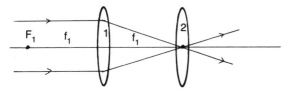

Fig. 4-17. Putting a second lens at the focal point of the first has no effect on the action of the first lens.

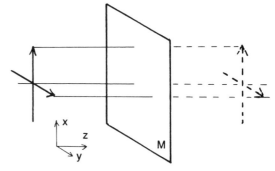

Fig. 4-18. The image in a mirror M has undergone inversion (but is not inverted).

or form an image on the other side of the second lens, again with no effect. This can also be seen by using $d = f_1$ in the preceding relation and finding the combined $f = f_1$.

4.9 PLANE MIRRORS

A flat mirror is made with a glass sheet with silver, aluminum, or other metal film deposited on the back surface. Ordinary plate glass seems smooth but is rough at the scale of visible wavelengths. For fine optical work, a higher degree of flatness is required, and the glass must be ground and polished flatter than a fraction of wavelength. Reflection also occurs from the front surface, giving a ghost image that can be annoying in everyday applications and intolerable in optical work. For fine mirrors, the fragile metal film is put on the front surface to make a "first surface" mirror. Aluminum is a good reflector from near ultraviolet to near infrared. Silver loses its reflectance just as the ultraviolet band begins. The term silvering often refers to any metal film put on glass. A partially silvered mirror has such a thin film that part of the light passes through it. It is inaccurately called a one-way mirror, as light can pass both ways. There are also ways to coat glass to suppress unwanted reflections, as will be seen.

Experiment 4-7

Look in a mirror. As you move your head around, does every part of the image seem to move, or just the image of your head? Test how it is that your right hand seems to be left in the mirror. Why is your head not upside-down? ■

The image of objects in a plane mirror consists of a mapping of points from object to image such that every point maps normal to the surface and is im-

aged the same distance behind as the object point is in front (Fig. 4-18). The image is not magnified but appears full scale. The mirror does not exchange right for left while leaving up and down alone. The image of each hand appears directly in front of the respective hand, just as for the head and feet. The mirror reverses one dimension, the dimension pointing at the mirror. This transformation is called *inversion* but is not to be confused with the action of a lens. Reversing one dimension changes any coordinate system into a nonequivalent one—for example, a right hand into a left hand. A large word referring to two forms related as mirror images is *enantiomorphs*.

Experiment 4-8

Look in a mirror while rotating it slightly. Does the scenery swing past more rapidly than the rotation rate? ■

Mirrors are important for instruments partly because a moving mirror can magnify greatly a small effect. For example, an experiment to measure gravity uses weights hanging on a fine but strong thread. A mirror is glued to the thread. The slight attraction between the weights and nearby fixed weights rotates the thread slightly. A light beam shining on the mirror and reflected across the room will swing through a large distance when the thread twists imperceptibly. Any turning of a mirror automatically doubles the arc through which the image swings, as found when trying to orient a mirror to see a certain region across the room.

Experiment 4-9

Stand between two large parallel mirrors. Can you align them to see an infinite number of images? Why do the "distant" images grow darker? ■

4.10 SPHERICAL AND OTHER CURVED MIRRORS

Mirrors with curved surfaces in a spherical shape are easy to grind, but astronomers prefer another shape. Distant starlight arrives as essentially parallel rays. The principle of least time can be used to determine the shape of reflector that will bring parallel rays to a focus F. In Fig. 4-19A, a line L is drawn perpendicular to the parallel rays. A mirror surface is then constructed (in cross section) with the requirement that any ray strikes the surface at point A such that AF = AL. This is exactly the definition of a parabola. It is fairly easy to show mathematically that the central region of a spherical mirror curves approximately like a parabola. (But astronomers prefer a fully parabolic mirror.)

A spherical mirror has a radius $R = 2f$, where f is the distance from mirror to focal point F on the axis. With this established, the same formulae that

worked for lenses work for mirrors, with the same sign conventions. A spherical concave mirror can produce real images if the object is not closer than F. Ray tracing also works almost the same as with lenses. As shown in Fig. 4-19B, a ray parallel with the axis is reflected through the focus. A ray aimed at the center of curvature C returns through it (the ray is on a diameter of a sphere). A ray aimed at the focus is reflected parallel with the axis. Thus the minimum of two rays needed to locate the image, and give its size, are found. All this works properly only near the axis and for rays whose angles are small with respect to the axis (exaggerations in the figures not withstanding). A car headlight has the small glowing filament at the focus of a low-quality spherical mirror. Since a slightly spreading beam is wanted anyway, the size of the filament prevents formation of a parallel beam of light.

The same ray and mathematical methods also work for convex mirrors that can produce only virtual images. These are rather familiar from supermarket aisle monitoring, where they are used to provide a wide angle view to a wide range of observers and with great clarity. The accompanying distortion is also familiar and unavoidable. A mirror

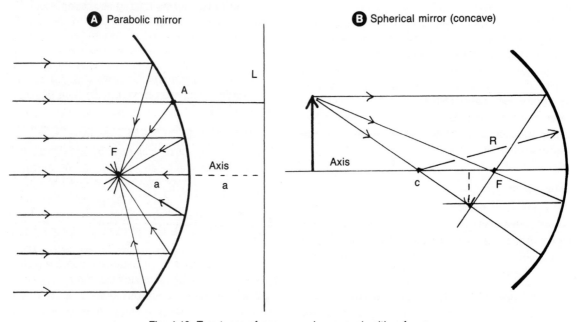

Fig. 4-19. Two types of concave mirrors, each with a focus.

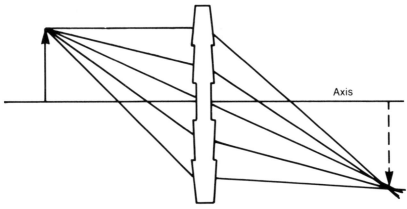

Fig. 4-20. A Fresnel lens (in cross section) is approximately flat
with specially refracting rings. (Only five are shown for simplicity.)

can be designed that captures all light from a source and focuses it in one spot. The requirement that the distance from one focus to the mirror to the other be constant is precisely the definition of an ellipse. A three-dimensional ellipsoidal mirror encloses the source. An advantage of curved mirrors for optical focusing is that several kinds of aberration of the image are eliminated. The light strikes only front surfaces with reflecting metal film. It cannot suffer dispersion or absorption.

Experiment 4-10

The reader familiar with art might locate some of the many cases where artists have shown the use of plane and curved mirrors in paintings—for example, art by M. C. Escher, Jan van Eyck, Edouard Manet. Are the laws of reflection and the proper orientation of images shown correctly in each case? Does the painting pass the test when you diagram the location of viewer, mirror, object, and painting?
■

4.11 FRESNEL LENSES

A different way to make a lens results in a thin lens with a close focal point that has as large a di-ameter as wanted. This method will not work with glass, which will be too fragile in large, thin sheets, but has been adapted to plastic. The principle uses the glass block model of a lens. The different parts of the lens are given slants at different angles as shown in Fig. 4-20. The outermost parts slant the most, and the innermost parts are nearly flat. Each section of the lens is formed as a ring, and all rings usually have the same width. The face of this Fresnel lens appears as if it were scored with grooves. This lens is easy to make by molding acrylic or other plastic. It cannot give a sharp image because those parts of the lens at the edges of the rings trap or send light rays in unplanned directions. But it is good enough for applications such as overhead projectors and focusing solar energy.

The principle of least time must not be absolutely necessary in using a lens to form an image. The Fresnel lens forms an image only by the law of refraction and does not equalize the travel times of all rays. The different rays from an object point take different times to reach the conjugate image point. The resulting degradation of image is subtle and usually unimportant, especially when compared to the errors introduced by the circular grooves in the lens.

5

Advanced

Geometric Optics

Mention has been made of the imperfections that appear when simple lenses and mirrors are used for optical purposes. As anyone knows who had inexpensive optical toys as a child, a toy telescope just does not perform well. More advanced optical methods are needed to correct the many aberrations and obtain the bright, sharp images that you are accustomed to in a high-technology society.

5.1 THICK LENSES

Thin lens analysis is not usually justified because most single lenses have substantial thickness and because advanced lens systems use compound lenses that are very thick. For a thin lens, it was assumed that all refraction of rays occurred in one plane. Thick lenses can be modeled with two focal points as before and with two planes for bending, called *principal planes*. Rays are bent at the principal planes.

More sophisticated terminology is also needed to discuss advanced geometric optics. A lens or lens system is also called a transform or transfer function and maps each object point into an image point.

These conjugate points are related in that light can travel either way, starting from one point and arriving at the conjugate point.

As shown in Fig. 5-1, a thick lens is modeled with two principal planes PP separated by a distance d. No symmetry can be assumed except that the focal points F and F' are each located distance f from the respective PPs. The PPs are not truly planar but are assumed to be, at the cost of small errors. For some shapes of lenses—a very wide variety is possible—the PPs are one or both outside the actual glass. But generally the separation d is about one-third the thickness of the glass.

The Gaussian, Newtonian, and magnification relations of Chapter 4 can be applied to thick lenses, provided that object distance x is measured from the front PP and image distance x' from the back PP. The focal length can be calculated easily from the radii of curvature and is given in the references (for example, Hecht & Zajac). The locations of the PPs can be calculated also, if needed. The only change in the rules for ray tracing is that any ray reaching

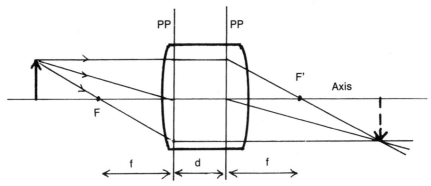

Fig. 5-1. A thick lens can be represented by two principal planes (PP) and the usual foci.

a front PP is carried straight across, parallel with the axis, to the back PP before continuing according to the rules. The location of an image is shown in Fig. 5-1.

5.2 MONOCHROMATIC ABERRATIONS

Monochromatic aberrations are errors or distortions of focus due to imperfect geometry, and they occur for any and all colors of light. Almost all would be unimportant if optical systems could use only paraxial rays very close to lens axes, but this is not practical. The reason for the many different shapes of converging and diverging lenses will be apparent. Some aberrations can be cured without going to special lenses. Some cannot be cured even if nonspherical lenses can be designed and made. To some extent, aberrations made by one lens can be canceled by a defect in another lens. The order in which the aberrations are discussed is also the general order in which they must be corrected in order to avoid worsening one while correcting another.

In *spherical aberration*, the rays that are farther from the axis are bent more and focused closer to the lens (Fig. 5-2). A spherical lens refracts too well near its edge. Cones of rays going through the lens focus in slightly different places, depending on how far from the axis they are, so that the focal point is stretched out into a long, thin region called a *caustic*. There is a position of best focus which depends on how stopped down the lens is. Obviously, stopping all rays except those near the center reduces

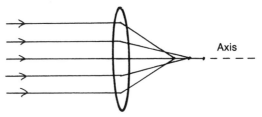

Fig. 5-2. Spherical aberration results in marginal rays being focused closer than axial rays.

spherical aberration. It also greatly reduces the brightness because most of the light passes through the outer parts of the lens. Another way to decrease spherical aberration is to use lens shapes such that incoming rays are bent about the same at the first surface as they are at the second surface. The basic lens relations have been refined to show the mathematical basis for spherical aberration. The focal length of the lens depends on the distance that rays are from the axis.

Experiment 5-1

Slightly tilt a lens while trying to focus the sun or a bright point source on a screen. What does the image resemble, and which way is it aimed? ■

Coma (from Latin for hair) is an aberration in which object points away from the axis are not imaged in one place or even in one plane. The PPs representing the lens are not planar either. Coma is readily observed by slightly tilting a lens while trying to focus a bright point source on a screen. The

51

image appears in the form of a small comet (from the same Latin word). Different cones of rays through the lens focus in different-size circles at different distances from the axis as shown in Fig. 5-3. In positive coma, the comet points toward the axis, and in negative coma it points away. Varying the shape of the converging lens varies the coma from positive through none to negative. There is usually a particular shape which, in combination with a particular object distance, gives negligible coma. A mathematical relation called the Abbe sine condition shows that coma is avoided when the lens is arranged to have incoming and outgoing rays at cer-

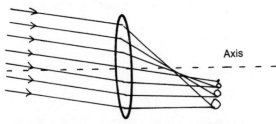

Fig. 5-3. Coma is an aberration where slanted rays have different focal points depending on which part of the lens they passed through. Positive coma is exaggerated here for clarity.

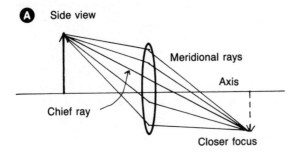

Fig. 5-4. An off-axis object suffers astigmatism because one set of rays passes asymmetrically through the lens and focuses closer than the other set.

tain angles and to have refraction occur at an imaginary spherical surface. A stop can be located in such a way as to let the central or chief ray of a bundle pass through the center of the stop, preserving symmetry for the cone of rays.

In *astigmatism* the rays from off-axis object points encounter different shapes of lenses in different planes and arrive at different focal points. Consider rays from the top of an object, shown in side view in Fig. 5-4A. These rays are in a *meridional* plane and pass through an asymmetrical part of the lens. Meanwhile, in top view (Fig. 5-4B) another set of rays from the same point are in a *sagittal* plane and strike the lens symmetrically. The focal points are separated for the two planes of rays.

A simple way to test for astigmatism is to use a test pattern made of dots as shown in Fig. 5-5. In the two different focal planes, (meridional and sagittal) there will be two different blurrings of the image of the pattern. In the meridional focal plane, the dots have blurred tangentially. In the sagittal focal plane, the dots have blurred radially, forming little *sagitta* (Latin for arrows) pointing toward the axis. These effects can be seen only if the lens is free of the preceding aberrations (spherical and coma). This astigmatism occurs for spherically symmetrical lenses. Astigmatism, when applied to flaws in the eye, is due to assymetry of the eye lens and has different results.

Petzval field curvature (named for Josef Petzval) occurs because the image is not formed in a plane (Fig. 5-6A). Only the center of the image would be in focus on a flat screen. However, object points located on a certain spherical surface would image to points on another certain spherical surface. For a positive (converging) lens, the image surface curves toward the lens; for a negative (diverging) lens, it curves away. Field curvature can be minimized (flattened) if thin negative and positive lenses are combined. Negative and positive lenses with the same focal length can be combined to give a useful net focal length if sufficient separation is put between the lenses. The requirement is shown mathematically in references on aberrations. Photography onto flat film and movie projection onto flat screens would be very disappointing unless Petzval

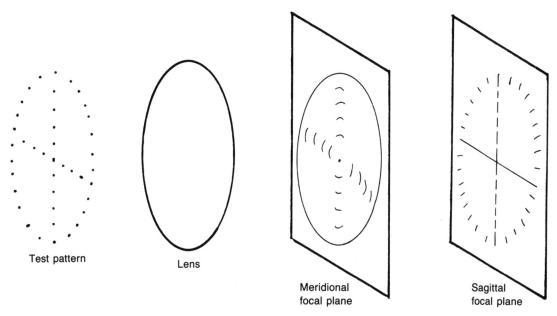

Test pattern   Lens

Meridional focal plane

Sagittal focal plane

Fig. 5-5. A simple test for astigmatism.

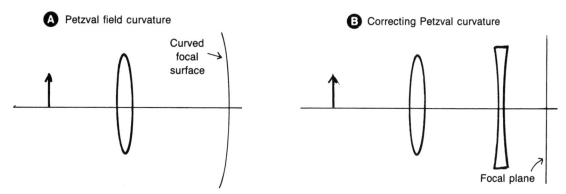

A Petzval field curvature

Curved focal surface

B Correcting Petzval curvature

Focal plane

Fig. 5-6. Petzval field curvature results in a curved focal surface.

aberration is corrected. This is often done by adding a negative field lens to the main lens system (Fig. 5-6B).

Distortion is due to the variation of lateral magnification with distance from the axis. The effect may be positive or negative, with parts of the image being stretched away from the axis or shrunk toward the axis. A good test pattern for distortion is graph paper as shown in Fig. 5-7. Positive distortion or pincushion means the whole image is larger than normal, especially its outer parts. Negative distortion

or barrel means the image is shrunk, especially its outer parts.

Thin lenses give little distortion, and thick ones give much. Positive lenses give positive distortion, and conversely for negative lenses. Distortion is minimal if the lens is used to give a magnification of unity, not always a useful arrangement. Stops not at the lens exaggerate distortion because they cause the effective object distance to vary for different parts of the object, and therefore the magnification varies. As shown in Fig. 5-8, the cone of rays around

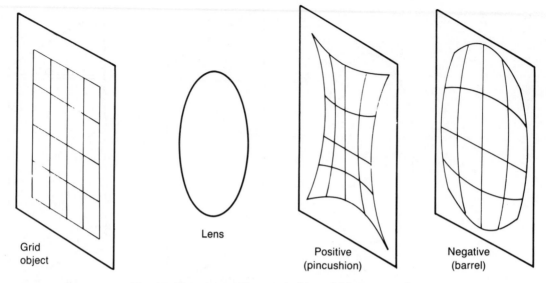

Fig. 5-7. Distortion can be tested with a grid (exaggerated).

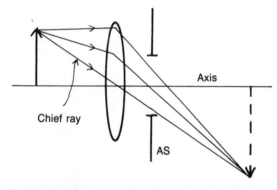

Fig. 5-8. A stop causes positive distortion because the chief ray is put at one side of a bundle of rays.

the chief ray that goes through the center of the stop travels less distance to the lens, equivalent to having a slightly closer object. The magnification is then more, and positive distortion results. For a similar reason, negative distortion results from a stop in front of the lens. A stop used halfway between two lenses would cancel the distortion from both lenses. This practice, along with using a magnification of unity, eliminates several other aberrations, too.

5.3 CHROMATIC ABERRATIONS

Chromatic abberration is due to different refraction by optical materials for different colors of light. For a poor, uncorrected lens, chromatic aberrations dominate the monochromatic ones. There are two types, axial and lateral. A converging lens focuses blue closer to red on the axis, with other colors in between (Fig. 5-9A). Off-axis points are focused to different heights as well as positions in different colors because of different magnifications (Fig. 5-9B). Negative (diverging lenses) show the opposite effects.

Experiment 5-2

The first problem you will notice when looking through a simple lens at a bright source is red or blue colored edges on images. Move the lens back and forth to see different colors emphasized. ■

The key to correcting chromatic aberration is to combine a positive lens with a particular index of refraction with a negative lens having a different index. The resulting combination is called an *achromat* and can cancel chromatic aberration for two colors. The colors to be corrected are chosen from near the ends of the visible spectrum (for example, blue and red), and some mismatch will occur in the middle (yellow). Using the simple lens relations and data on the indices of refraction for the chosen colors, the focal lengths and types of glass for the pair of lenses can be chosen.

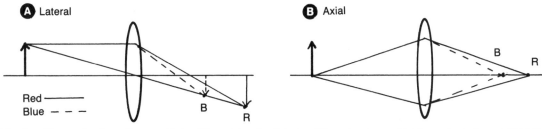

Red ——————
Blue — — —

B

R

B

R

Fig. 5-9. Chromatic aberration results in different focal points for different colors, with blue refracted more than red.

Fig. 5-10. An achromat is a compound lens of two glasses to correct chromatic aberration at two wavelengths.

The information available on the many different kinds of glass, whose indices range from 1.4 to 2.0, is specified in terms of performance at precisely measured wavelengths. These colors or wavelengths are chosen to be spectral lines commonly available with laboratory equipment. (Spectral lines will be explained further in Chapter 12.) The relations of the indices for two or three common spectral lines (usually blue and red wavelengths from hydrogen, and yellow from sodium) are combined to form design numbers called Abbe numbers (named for Ernst Abbe), which represent the amount of dispersion for each kind of glass.

The most common achromat is the Fraunhofer cemented (Fig. 5-10). A double convex lens of crown glass (low index) is cemented to a plano-concave lens of flint glass (high index). This combination can be designed to correct spherical aberration and coma also. If the achromat is not too thick, both types of chromatic aberration are corrected. Otherwise, separated pairs of achromats are needed to truly reduce chromatic aberration. It is possible to combine a selected polished mineral lens with glass to form achromats that correct at three different colors. Or combinations of three or more glass lenses can correct at three or four colors. The cost rises rapidly.

5.4 THE EYE, ABERRATIONS, AND CORRECTIONS

The eye is a complex optical system, although it has just two sophisticated lenses. Technology has not yet perfected a way to change the shape of a lens the way the eye does in order to change focus while minimizing aberrations. This discussion of the optics of the eye is a physical introduction. The implications for visual perception are left to Chapter 16.

Like a single lens, the eye lens forms an inverted real image. Ideally it is focused on a curved screen called the retina enclosed in a lightproof chamber. The spherically curved focal plane is matched to reduce aberrations produced by the lens. As shown in schematic form in Fig. 5-11, the eye lens system has two parts, the cornea and the crystalline lens. The latter is not well-named because the lens is flexible and will henceforth be called the lens. The cornea is somewhat flatter than spherical to reduce spherical aberration. The cornea has a focal length of about 16 millimeters, but no one can see this closely. Between the two lenses, where it belongs, is a variable stop or diaphragm, the iris (the colored part of the eye). The aperture or pupil can vary from 2 millimeters to 8 millimeters in diameter. Since the best focusing is found through the centers of most lenses, the eye can form the best images when the pupil is small. The pupil can control light intensity over a ratio of 1 to 16.

The largest change in index of refraction occurs at the air-cornea boundary. All inner parts are liquid filled and have indices near that of the cornea—

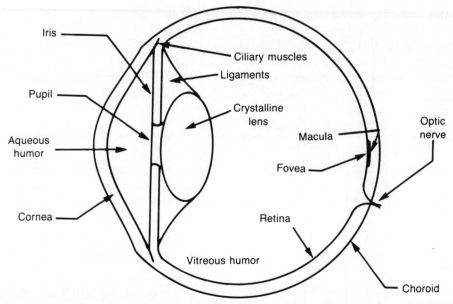

Fig. 5-11. A cross-sectional representation of the optical structure of the human eye (no scale).

and near that of water (1.33). The cornea has an index of 1.38 and the aqueous humor, between cornea and lens, has an index of 1.34. The lens has complex layered structure with an index of 1.41 in the center and 1.39 at the edges. Artificial lenses with variable index are still a technological hurdle. Between this lens and the retina is the vitreous humor with an index of 1.34.

Experiment 5-3

Squint at a brightly lit area and you can see the outlines from bits of cellular debris floating in the vitreous humor. ■

The lens shape can be changed by small ciliary muscles, so that its focal length is variable to allow focusing of far and near objects on the retina. Another important component of optical systems, dull black surfaces to absorb stray light, is found as the dark-pigmented choroid lining the eye all around (except at the cornea). On its inner surface is the vital retina, covered with cells sensitive to light intensity and color. The rod cells are large, cylindrical cells very sensitive to light intensity but ignorant of color. The cone cells are smaller, less densely arranged (except at the macula and fovea), and are responsive to higher light levels. They are able to distin-

guish three colors—red, green, and blue.

As will be appreciated in later discussions of optical information, the eye has an information problem. To see a view of 180° or so in fine detail would require an unrealistically large number of cone cells. The ten million or so cone cells are not sufficient. Instead of more cells, there is a small region called the fovea densely packed with small cone cells where fine details of the image can be discriminated. The fovea is positioned so that when the eye is aimed most directly at an object, the image falls on the fine-grained part of the retina. Large objects must be scanned to see fine details, and the abrupt natural eye movements are called saccades.

When the eye muscles are relaxed, the lens is flattest and has the longest focal length. Distant objects are then focused sharpest. Because the distance between lens and retina does not change, nearer objects must be focused by reducing the focal length of the lens so as to keep the image distance constant. The process of changing the focal length is called accommodation. We seem to have greater depth of field than we actually have because the lens responds so rapidly to changes in object distance as we examine a scene. The closest objects that can be focused vary from ten to a hundred cen-

56

timeters away, depending on age. As it ages, the lens is less easily altered, and close objects can no longer be seen clearly. When designing optical instruments, images must be produced that the eye can comfortably focus. They should be as distant as possible. The eye can also see virtual images whose rays diverge.

The earliest use of artificial lenses was to correct problems in human vision, at least as early as the 13th century in Europe and earlier in China. The concept of the power of a lens was developed first for correcting vision. It is the reciprocal of the focal length. Opthamologists once had cases of lenses of different focusing power and searched for the one that made vision clearest. Now variable power instruments are in use. The normal eye has a power of about 59 diopters, with the cornea providing about 75% of this. The lens proper is a fine focusing device.

The problem in nearsightedness or *myopia* is that distant objects are brought to a focus in front of the retina. The lens system is too strong (Fig. 5-12A) or the eye is too long. There is a ''far point'' beyond which objects cannot be imaged properly. Myopia is corrected with a negative lens as shown in Fig. 5-12B. The lens causes rays from distant ob-

jects to diverge sufficiently that the eye lens can accommodate them. The added negative lens can be wrapped around the eye somewhat as in modern spectacles, so long as its thickness varies in a concave way. The correcting lens is located at the focus of the cornea so that it has no effect on the magnification of the eye lens system. Since the eye lens loses accommodating power with age, nearsighted people often find their sight improving as they grow older; then it may shift to farsightedness.

In farsightedness or hyperopia, the eye lens system is too weak, and distant objects are focused beyond the retina (Fig. 5-12C). Or the eye is too short. There is a ''near point,'' inside of which objects cannot be imaged properly. Hyperopia is corrected with a positive lens in front of the cornea so that rays from nearby objects are caused to converge more strongly (Fig. 5-12D). Again the magnification of the overall system is not changed.

Many eyes suffer from astigmatism, which in this case is due to an asymmetry of the cornea and not just due to rays passing through it asymmetrically. As discussed earlier, astigmatism affects the quality of the image overall and would not cause just a distortion of parts of it. The cornea can be asymmetric in many ways, and sometimes correcting

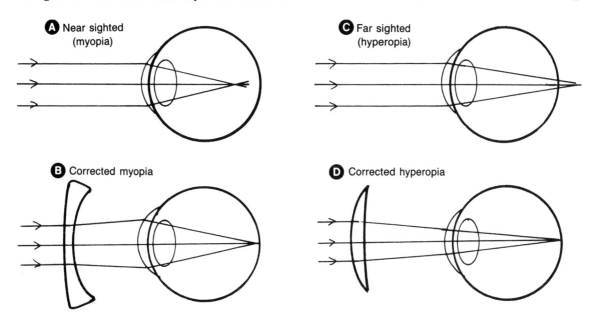

Fig. 5-12. Correcting nearsightedness (myopia) and farsightedness (hyperopia).

lenses cannot be designed to correct it. When it can be corrected, an anamorphic lens is needed with different power in two perpendicular planes. A slice through the side of a barrel is a model of one useful shape that has different spherical radii in two different planes.

There are ways to focus other than by deforming the lens to change focal length. Eyes have developed separately in many species of animals, and nature has explored every optical option. Some animals can move the lens back and forth with respect to the retina to obtain focused images. Some birds can change the curvature of the cornea, which can be rapid and effective because the refraction from air to cornea is much larger than from lens to vitreous humor.

Experiment 5-4

If you swim underwater, examine the ability of the eye to focus underwater. What is hindering it, besides irritation of the eye? ∎

5.6 ADVANCED MATRIX LENS ANALYSIS AND DESIGN

The modern approach to optical analysis is to arrange for efficient computer calculation. Ray tracing is still done to test lens systems. But the simple rules involving parallel rays or rays through foci cannot be used because they do not reproduce the full effects of aberrations. Instead, a chosen paraxial ray is followed through each optical surface, wherever the ray may go according to the law of refraction. This was often done in the laboratory with actual lenses and high intensity spectral lines or colored lasers. More recently the lenses are modeled mathematically, still assuming spherical surfaces. The number of calculations is huge and tedious, but computers can check any complex system in minutes. It is still up to human optical engineers to choose which lenses to insert where and to move them around in search of a minimum in all aberrations of concern. The benefits of these advances reached the marketplace in the mid-1970s as superior camera lenses became available at much lower prices.

To trace a light ray of a particular wavelength through a series of lenses at known spacings and made of certain materials, Snell's law is applied at each surface where the light ray passes through. The series of calculations can be done as a series of matrix multiplications. A matrix is a set of numbers that represent the conversion of two or more known quantities into two or more new quantities. The location (in two dimensions) and direction of the incoming ray is known. A refraction operation is done on it at the first surface by means of a matrix multiplication. Then a transfer operation is done to find where the ray strikes the next surface. This continues through air and various glasses in the order in which the ray encounters the materials until the outgoing ray is found. When many suitable test rays have been traced, the image points are plotted to show what sorts of aberrations remain in the lens system, as well as to find the position and magnification of the image. Skew rays, not traveling in a plane containing the axis, were very difficult to test until this method appeared.

Since the molding of plastic lenses and mirrors with high precision is now possible, the design of nonspherical surfaces is a new option for avoiding some aberrations. In effect lenses are designed as Fermat's principle would have them, bringing a wide cone of rays from a wide range of object points to a focus by making each ray travel in the same time. The corresponding lens molds are then made, and superior optics can be turned out at competitive costs. Plastics do introduce a thermal expansion problem that glass rarely caused, because most plastics vary their dimensions up to ten times more than glasses as temperature changes.

As the next chapter shows, modern applications require a large number of different and complex lens designs that minimize the seven aberrations. Besides computerized modeling, other advances have allowed faster and cheaper designs with higher quality. Those advances include materials with higher indices of refraction (some minerals are being used with n well over two) and new ways to eliminate reflections from lens surfaces so that lenses need not be cemented together. The measure of quality centers around resolution and, more recently, spatial frequency bandpass—both later topics.

6
Optical Instruments— Applications of Geometric Optics

You have now read sufficient optical theory that many common optical devices and instruments can be described in some detail. Geometric or ray tracing principles are sufficient to allow design of most aspects of these devices and to allow understanding of their functions (with some exceptions). The oldest optical device, the correcting eyeglass, has already been discussed. Eyeglasses are not intended to magnify, only to assist focus.

Optical instruments well outside the visible range and electron microscopes, which use particles, are not included in this book.

6.1 MAGNIFIERS

The simplest optical device is the magnifying lens. In order to examine a small object (one of the major applications of optics) it can be brought closer to the eye, up to a limit of about 25 centimeters. To make it seem closer, a positive lens is needed to make the rays converge more (Fig. 6-1A). The magnifier is used closer to the eye than its focus so that the image is uninverted and therefore virtual.

Any other arrangement results in smaller, real, and inverted images, not what most people can adjust to in daily work.

To explain the magnifying ability of this and more complex instruments, the concepts of angular size and angular magnification are needed. As shown in Fig. 6-1B, the angular size of an object can be defined approximately as the ratio of height y to distance x. The triangle shown defines the *angular size* such that:

$$\tan \alpha = y/x$$

The *tan* (tangent) function of the angle can be approximated by the angle itself when the ratio and the angle are small (as often occurs in lens optics), so that $\alpha \cong y/x$. The angular size α is measured in radians, a unit that results naturally from calculating an angle from a ratio. (2π radians make up $360°$, so one radian is approximately $57°$.)

The net effect of an optical instrument is to change the angular size α of the object to a larger

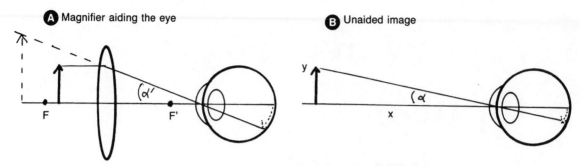

Fig. 6-1. A lens as magnifier to be used with the eye.

angular size α' of the image as seen by the eye. The *angular magnification* is defined by:

$$M = \alpha'/\alpha$$

This magnification, also called magnifying power, is not the same as lateral magnification because the effect on the eye is included. The reason for defining angular magnification for a lens used in this way is that the lateral magnification must be zero for an object or image at infinity. The virtual image formed may be infinitely big and infinitely far, but the ratio of lateral size to distance is quite definite and constitutes a definite angular size. The angular magnification depends on the relative locations of object, lens, and eye. The angular sizes to be compared are usually understood to correspond to the object at the *near point* of best vision (25 centimeters from the eye) and the virtual image of the object seen in the magnifier.

If the magnifier is placed so that its focal point is at the object, the virtual image will seem to be at infinity. Parallel rays are sent to the eye so it can relax while examining the image, and the location of the lens is not critical. Analysis shows that the angular magnification is $25/f$, where f must be in centimeters. If the magnifier is put slightly closer to the object (Fig. 6-1A), the virtual image is formed nearer, preferably about 25 centimeters away where the eye has the most distinct vision. Now the eye must accommodate diverging rays, and it matters more where the lens is placed. The user will tend to hold the lens close but not so close that viewing the image is uncomfortable. Magnifiers permit lati-

tude in placement. Analysis shows that one is added to the previous magnifying power.

Greater magnification is obtained from a lens of shorter focal length. When magnification over 10 is achieved, the focal length must be less than 2 centimeters. This is a highly curved lens. The radius of curvature is the same as the focal length, so it can only have a small diameter typically less than half its focal length. Therefore, a reading magnifier that will take in more than one word at a time cannot have high magnification because it must have a large size and long focal length. A large, high-power magnifier is impossible. At higher powers, aberrations are a serious problem, otherwise a simple glass marble would be the strongest magnifier. Images with low aberration are obtained with compound and achromat magnifying lenses.

6.2 EYEPIECES

A magnifier can be used with real or virtual images formed by another optical instrument such as microscope or telescope. In this case, it is called an *eyepiece* or *ocular*. The eyepiece accommodates the instrument to the eye. If the instrument produces a real intermediate image, the eye cannot see it, just as a focused image in midair cannot be seen. The eyepiece must provide a virtual image. Rather than design the entire telescope, for example, as one system, it might be designed to produce the best intermediate image. That image might be passed on to a camera with an adapting lens system or to the eye through an eyepiece. The goal is to produce a virtual image near infinity so that the eye can relax. The rays from any image point will be parallel.

The eyepiece should also not have to be placed directly against the eye. The eye should be put at an imaginary "eye point" behind the eyepiece. Here the exit pupil of the eyepiece should match the pupil of the eye so that light is not lost. An expensive instrument that concentrates dim light for the eye would be a waste if most of the light landed outside the opening of the pupil. It may seem simple to use one lens to match an image to the eye, but any practical attempt will show that much care has gone into designing good working eyepieces with low aberrations. There are a limited number of designs, and some are interchangeable on some instruments. The analysis of magnification is similar to that for magnifiers, when the effective focal length is known.

An old and common design is the Huygens eyepiece (Fig. 6-2A). Like most, it has two lenses, the input or field lens and the eye lens near the eye. It is designed to accept a virtual image. A stop is placed where the virtual image is formed to limit extraneous rays and define the exit pupil. Cross hairs or a graticule (a grid or other precision marks on a glass plate) can be placed there and viewed along with the image. The eye lens renders the rays parallel and is forced to be very close to the eye. The eye sees the image, the graticule, and the exit pupil all together.

An improved design is the Kellner eyepiece (Fig. 6-2B), which requires the virtual image to form in front of the field lens where the stop is placed. The eye is put farther from the eye lens, and an achromat corrects chromatic aberration.

Other eyepieces include the orthoscopic wide field eyepiece, the Erfle with three achromats and a wide-field (widely used and expensive), the old Ramsden, and the Barlow lens.

6.3 MICROSCOPES

To see smaller nearby objects than a magnifier will permit, many sorts of microscopes have been invented. These are limited in magnifying power by aberrations and by the wavelength of light itself (for reasons given in Chapter 7). When the object is smaller than the wavelength of violet light, there is no hope of focusing on it with visible light. Light microscopes have been developed to nearly this limit. The first one was probably made in the late 16th century in Holland and resembled the compound microscope of Fig. 6-3. The lens near the object is the *objective*, with a short focal length. Its diameter is as large as possible so it can gather the most light, but with a short focal length, it must be a small lens. The object is placed just outside the focal point so that large magnification is obtained for

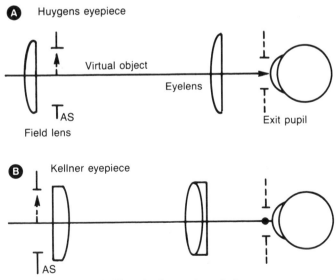

Fig. 6-2. Two simple eyepiece designs.

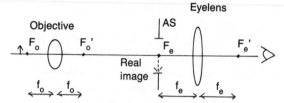

Fig. 6-3. Compound microscope.

the real image. A field stop at that position limits rays to the largest the lenses can handle. This real image serves as object for the one-lens eyepiece (eyelens) and is located at the focal point of the eyepiece. Essentially parallel rays are sent to the eye so that the image is seen virtual and inverted. It is magnified in angular size by some power.

The overall magnification is the product of the lateral magnification of the first lens and the angular magnification of the second treated as a magnifier. Typical magnifications are in the range 10 to 100. Aberrations are aggravating unless compound and achromat lenses are used, and these improvements are essential at higher powers. Focusing for different object positions and sizes is done by moving the barrel carrying both lenses slightly up or down. If you have looked into a microscope, the exit pupil is very apparent and you have naturally moved your eye sufficiently close so that you could see all of the field available. High magnification without spherical aberration can be achieved with an *oil-immersion* objective. An oil with the same index of refraction as the objective makes contact between specimen and lens.

Stereoscopic microscopy is achieved by providing an eyepiece for each eye. Only one objective is needed because the two points of view are so close at the tiny object that the same lens can be shared. The highest achievable magnification is somewhat over 1000. Since the eye can distinguish angular sizes of about 1 minute of arc, the best light microscopes can show detail down to about 1/16 second of arc. For the working distance of the eye (25 centimeters), this translates to about 1 ten-millionth of a meter, or less than the wavelength of light. Therefore the highest power microscopes are valuable for brightness rather than for the finest detail.

The background or *field* of the image in a microscope can be light or dark. When the specimen is illuminated from below, direct and scattered light travels through the microscope, giving a light field. When the specimen is illuminated by light that strikes it but not the objective, then only light scattered from the specimen passes through the microscope, and the field is dark. When the specimen is thick and the magnification high, the complex image is confusing. The illumination can be focused on one plane of the specimen, and this can be varied, so that the image is a scan through the specimen. This is *confocal* microscopy.

Seeing a layer of the specimen has been difficult until the recent invention of the tandem scanning reflected light microscope (TSRLM). In this instrument, (see references) incoming light passes through tiny holes in a spinning disk. Then it is focused on some layer of a specimen. The reflected light returns the same way but is focused through different but corresponding holes and sent to the eyepiece. Light from other layers in the specimen cannot find a path through a viewing hole. The image is actually composed of numerous transient spots seen through the tens of thousands of holes involved, but it appears sharp and stationary. The method permits making stereoscopic pictures of specimens in depth, too.

6.4 TELESCOPES

Telescopes come in a wide variety. Most invert the image and are used for astronomical purposes where scientists prefer to work with the incorrect orientation of images rather than purchase an inverting system that is difficult to design, takes away more light, and brings in new aberrations. Often film is used to capture astronomical images anyway. Terrestrial telescopes are very inconvenient unless the image is erected somehow. Binoculars, which are a pair of short telescopes, are discussed in the next section and use a special method for reinverting. All telescopes are understood to be intended for viewing very distant objects that give parallel rays, although many can be adjusted for much closer objects. It can be argued that an infinitely distant object must be too small—not to mention too dim—to ever be magnified to visibility. Stars are a more realistic example. Their brightness can be increased with a telescope, yet their size remains infinitesimal (pointlike) with present technology, and their rays are parallel. A distant galaxy shows size through a telescope, so its rays do not arrive parallel. For design purposes, however, both stars and galaxies are assumed to be at infinity.

The earliest maker of a telescope, probably in the 16th century, is uncertain, but the type was surely a *refracting* one. This type uses lenses to collect and refract the light from a distant object so that the eye can see it magnified. The Dutch government bought the rights to an early design in order to develop it for military use in 1608. The next year Galileo, knowing of the idea, made a two-lens refractor. An early and unsatisfactory form called the Galilean telescope is shown in Fig. 6-4; it uses a convex lens as objective and a concave lens to convert

the real image to a larger and virtual one for viewing. Galileo soon improved upon this approach.

The refracting telescope shown in Fig. 6-5 has a large objective lens to gather as much light as possible and a smaller eyepiece lens to change the real image from the objective to a virtual one. A very distant object is imaged at the focal point of the objective f_o from the lens. The image of a star would be infinitesimal were it not blurred to a finite size by aberrations. However, more light has been gathered than the eye can gather, so the image is brighter. The eyepiece is a magnifier with focal length f_e placed so that the real image is at or just inside the focal point. Then the eye receives quasi-parallel rays from a magnified virtual image at a distance, just as in preceding discussion. A close object will form an image beyond f_o, but the eyepiece can be slid to view it magnified. The distance between the two lenses is close to $f_o + f_e$. This telescope also works as beam expander and shrinker.

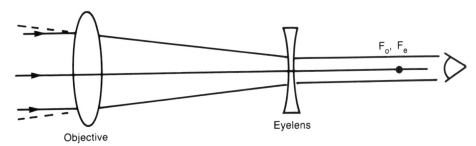

Fig. 6-4. Early Galilean telescope.

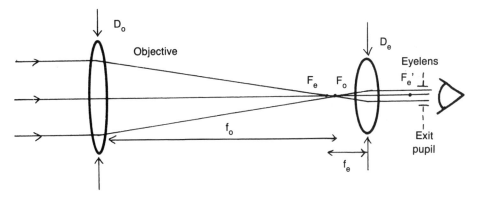

Fig. 6-5. Two-lens refracting telescope.

A wide beam of parallel rays shone into the objective is converted to a narrow beam exiting the eyepiece, and vice versa.

The entrance pupil has the diameter of the objective D_o. Its image is the exit pupil with diameter D_e. The magnifying power of this two-lens telescope is found to be:

$$M = f_o/f_e = D_o/D_e$$

If the shortest convenient eye lens has $f_e = 2$ centimeters then a magnification of 100 requires the objective to have $f_o = 200$ centimeters. Clearly the telescope is too long for hand use. The image quality is also poor at this magnification unless correcting lenses are used with the objective and the eyepiece is of special design. For terrestrial use, the image needs to be inverted again with one or a set of erecting lenses placed between objective and eyepiece. This lengthens the telescope tube, making short binoculars more useful. Chromatic aberration is so important in terrestrial use that the objective must be an achromat (which is difficult to make large). The largest lens objective in use is 100 centimeters (40 inches) at the Yerkes observatory in Wisconsin. Because its area is about 15,000 times the area of the eye pupil in the dark, it can reveal objects 15,000 times dimmer than the eye can see.

The magnification is limited by the eye pupil to 100/0.8 or about 125, not sufficient to show the disks of stars.

A *reflecting* telescope uses a large concave mirror to exceed the light gathering capacity of the largest lenses. The mirror is ground with a parabolic shape for most astronomical uses because this shape brings parallel rays to the best focus. The art of making single large mirrors reached a climax in 1949 with the 5-meter (200-inch) Hale telescope at Mt. Palomar. For almost three decades following, it was thought too difficult to make larger mirrors. Then new technologies for lighter-weight, single-mirror telescopes and for multiple-mirror telescopes were developed. A 6-meter telescope recently went into operation in the Soviet Union. The laser and computer are now the key tools for planning larger mirrors.

Mirrors in astronomical telescopes have the objective far from the focal point, so that only real images are produced. These are quite useful for photography, and some telescopes can be used without lenses, keeping aberration very low. Figure 6-6 shows several ways to reach the image formed at the focus. The best image is at the prime focus (Fig. 6-6A), but it is very inaccessible. The Hale telescope at Mt. Palomar is large enough so that an observer

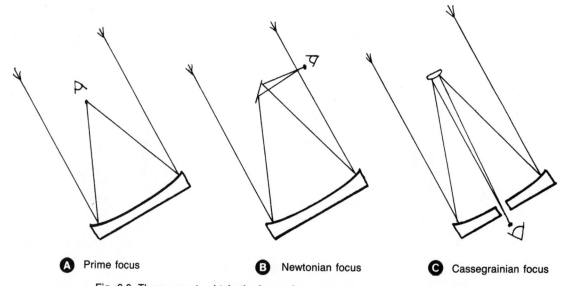

A   Prime focus   B   Newtonian focus   C   Cassegrainian focus

Fig. 6-6. Three ways to obtain the image from a concave mirror in a telescope.

can sit there without blocking too much light, but a photographic plate is best put there. Supports for equipment at the focus block little light and hardly affect the image.

The Newtonian arrangement (Fig. 6-6B) was developed by Isaac Newton in 1668. The focused rays are reflected to the side by a plane mirror. This puts the observer high on the side of a large telescope, which is an awkward arrangement. The image can be obtained beneath the big mirror if a hole is put in it as in Fig. 6-6C—the Cassegrainian form. Because the rays converge too sharply to exit through a small hole, a diverging mirror is put before the prime focus to increase the focal length effectively. In all arrangements, the image is real and ready for photography. If parallel rays are needed for visual examination or passage through instruments, a high-quality eyepiece is needed.

The wide field surveys of the sky were made possible by the development of the Schmidt *catadioptric* telescope. This term merely means that a combination of mirror and lens is used to form the first and real image. A low-cost spherical mirror is used, with a large correcting plate over it to precorrect spherical aberration (the only significant aberration from spherical mirrors). The corrector has special nonspherical surfaces. A parabolic mirror is not possible because of the wide field and resulting nonparallel rays. The magnifying power is low if this telescope covers a wide field. The film is put at the focus. Large film must be shaped to the spherical shape of the focal surface. The aperture is up to 120 centimeters (48 inches) and the f-number is 2.5. Other catadioptric telescopes have been developed for specialized nonastronomical observation.

The astronomer, wanting to gather the most light from dim objects, is usually interested in aperture rather than magnification, although high power accompanies large apertures if the focal length is long. Also important is the resolving power of the telescope, so that very close stars can be distinguished. The Earth's atmosphere varies in density, so that different parts of the air refract light slightly differently. This becomes more important when the mirror covers a larger area. The theoretical limit of resolution of about 0.03 second of arc is equivalent to about one part in 10^8 or a 5-meter object on the moon. The atmosphere is never calm enough to reach this resolution. Putting telescopes in space is one solution. Another is the use of laser calibration and computer-controlled correction to adjust mirror segments to cancel atmospheric refraction. The ultimate limit on telescope performance is, again, the wavelength of light. Space telescopes can easily approach that limit.

6.5 PRISMS—THEORY AND APPLICATIONS

Another kind of refractive and reflective device must be studied before certain instruments can be discussed. As shown in Fig. 6-7, a prism is a wedge

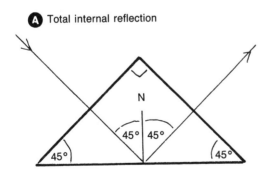

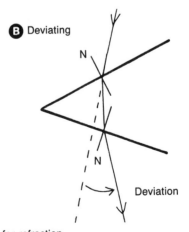

Fig. 6-7. Prisms used for internal reflection and for refraction.

of optical material that can either refract or totally reflect light, depending on the angle of incidence. The 45° glass prism shown in Fig. 6-7A is especially useful because light entering normal to one face will totally reflect out the other face, having changed direction 90°. Total reflection occurs because the light strikes the inner surface at 45°, which is greater than the critical angle of about 41° for glass.

Light striking an outer surface at an angle (see Fig. 6-7B) is refracted inside, reflected in part by any internal surface, and refracted again as it emerges. It has deviated from its original direction to emerge at a new angle (calculated in the advanced references). The general result is that the ray is bent partly back in the direction from which it came. The deviation depends on the angle in that vertex of the prism. For a symmetrical arrangement of incident and departing rays, the angle of deviation is a minimum—an important result.

Prism refraction has been useful since at least Newton's time because the variation of light speed with wavelength results in a separation of wavelengths of white light. Each different color emerges in a slightly different direction, with violet deviated the most. Because of the larger changes in the index of refraction at shorter wavelengths, blue and violet are spread more than red. Far away in a dark room, the resulting spectrum of white light can be examined in detail. If the entering light is very pure (one color only), no further separation occurs; the ray is simply bent toward the place it would go as if it were part of white light. Isaac Newton was the first (1672) to explore this dispersion of light with a prism and to show that a second prism would not break down an individual color further.

The prism was adapted in the early 19th century to an important instrument, the *spectroscope*, which shows the visible spectrum in fine detail. As shown in Fig. 6-8, a narrow slit and a lens (or collimating telescope) convert the light from any source to be investigated into a beam of parallel rays. The beam need not be narrow as it passes through a wedge of glass. The light should enter and leave the prism at similar angles for the largest dispersion. The rays leaving the prism are parallel for each color but must be collected by a lens before being examined directly by the eye or photographed. Because the dispersed light from a source is dim when spread over the spectrum, the whole apparatus is enclosed in a light-proof chamber. For work in the infrared or ultraviolet, prisms made of certain crystals are more effective refractors. Then the results cannot be seen directly but must be photographed or electronically sensed. The spectroscope enabled much progress in the study of atoms and stars. Some further details are given in Chapter 12. The prism spectroscope has been largely supplanted by a better device based on wave properties of light.

Prisms have many other uses. One of the simpler applications is the inversion of an image. Since internal reflection reverses one dimension of the image, two different reflections might reverse two dimensions, inverting it. A Porro prism (Fig. 6-9) is a 45°-90°-45° prism oriented so that light enters and leaves normal to the hypotenuse side. The letter R is selected here as an object because it has no symmetry and any orientation or reversal can be distinguished. The first two reflections shown serve to turn the image upside down without exchanging left and right. The image of R is then mirror-

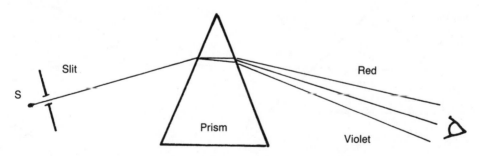

Fig. 6-8. Top view of a simple spectroscope using a slit and prism.

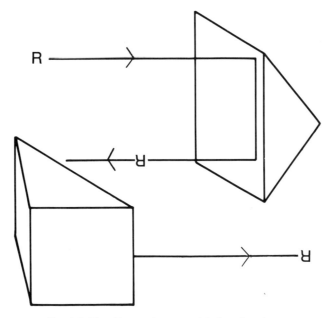

Fig. 6-9. Two Porro prisms used to invert an image.

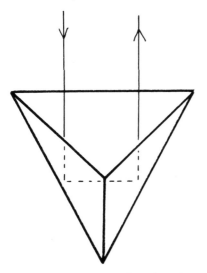

Fig. 6-10. A corner reflector (retroreflector) sends a ray back in exactly its original direction.

backwards and therefore not yet inverted. Another prism oriented sideways as shown reverses the other dimension of the image, producing a truly inverted R. Since nearly perfect reflection is used, the image is inverted without acquiring any of the aberrations a lens would cause.

More complex prisms have been invented that perform more complex changes in image orientation, simply by reflections. One of the most impressive is the corner-cube prism, made as if it were a slice off the corner of a glass block (Fig. 6-10). It has the geometric property of sending light back exactly in the direction it came. An array of these is now on the Moon to "retroreflect" laser beams directed there. The beam returns down the telescope used to send it. A mundane application of this principle is in tiny glass beads used in paint for highway signs that return headlamp light to drivers. Another use is prismatic plastic reflectors to provide passive night indicators of cars, mailboxes, and posts.

Prisms provide a very accurate way to measure index of refraction. A prism is made of the unknown transparent material and the deviation in direction is measured for any precise colors of light of interest passing through. The index is calculated from this data.

6.6 BINOCULARS

Not only does a set of crosswise Porro prisms reinvert an image without aberrations, it also adds considerably to the length of the path of the light

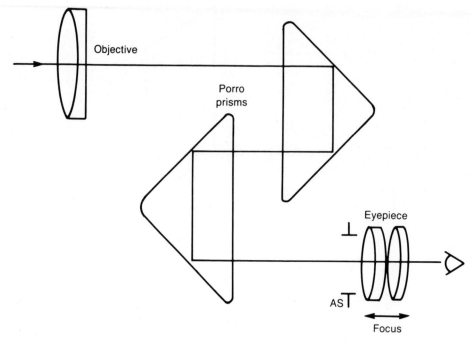

Objective

Porro prisms

Eyepiece

⊥

AS⊤

Focus

Fig. 6-11. A monocular employing two Porro prisms and achromats.

rays. This is an advantage when a telescope with long focal length objective is wanted short. In Fig. 6-11 a *monocular* (one half of a binocular) is assembled from a large objective lens (a quality achromat), a pair of Porro prisms to provide more length internally, and a complex eyepiece. The diameter of the aperture, when imaged, sets the exit pupil diameter, as usual. Typical magnifications are 5 to 20. Typical objective sizes are 30 to 50 millimeters. The optical specification is in the form 7×35, where 7 is the power and 35 is the objective. In this example the exit pupil is 5 millimeters, which is good for night viewing.

Experiment 6-1

Point a binoculars focused for infinity at a bright area such as the sky. Hold a screen in front of an eyepiece and notice the small, sharp-edged circle of light. This is the exit pupil. Measure its diameter and compare it with objective diameter divided by magnification. If exit pupil size does not vary with distance from the eyepiece, what does that say about the location of the virtual image you would see in the eyepiece? ■

To some extent, aperture and magnification can be varied separately—hence a wide variety of types of binoculars. But low magnification is likely to mean wide field, and vice versa. A large aperture does not necessarily mean the field will be large, but it does mean that more light will be gathered. Depth of field depends mainly on magnification—the higher the power, the more shallow the depth of focused field. The brightness of the image is proportional to the area of the objective. The brightness of a unit area of the image is inversely proportional to the magnification, because magnifying spreads out the light in the image.

Focusing the pair of monoculars that make the binocular is done by moving both eyepieces in or out simultaneously with a thumb screw. Good binoculars have individual focusing adjustments so that each side can be matched to that eye. The use of binoculars rather than a single telescope is meant to preserve the stereoscopic vision that results from looking at the same object with separated eyes. Binocular objectives are more widely separated and enhance the effect. Good ones permit the brain to readily merge the two images into one and detect the distance information contained in the combination of two images, thus seeing the object in depth. Stereoptic perception is covered further in Chapter 16. Binoculars should not be confused with field glasses, which use short telescopes containing erecting lenses instead of prisms and suffer from more aberration. However, some medium quality binoculars are made with complex "roof" prisms that do invert the image in-line. These are more compact and lack the offset light paths characteristic of Porro prism binoculars.

6.7 PROJECTORS

Projection lenses, widely used, are rarely discussed in optics books. They are similar to telescopes used in reverse. With a projector, a close bright object is transformed into a distant, real image on a screen. The object must be inverted so that the image is erect. The object, such as an illuminated slide, movie film, or small television screen is near the focus of a quality lens system (a single lens is shown in Fig. 6-12). A magnified image is produced at the other conjugate point where a screen is located. The source of illumination, being far from the front focal plane, is not magnified as an image. The light source also has a lens system, however, to form the rays from a small bright lamp filament into a uniform beam of parallel rays that strike the film. Often a spherical mirror is built into the bulb and set behind so that light going in the wrong direction is captured and sent forward.

A common projector lens design is the Cooke triplet (from the company of that name, in 1893). As shown in Fig. 6-13, two convex lenses are symmetrically placed about a concave lens of the same glass. This combination corrects all seven aberra-

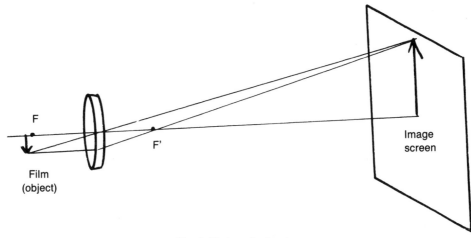

Fig. 6-12. A projection lens.

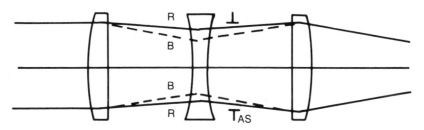

Fig. 6-13. The Cooke triplet lens eliminates all aberrations well.

tions including chromatic. A typical f-number for a slide projector is 2.5. For a typical exit aperture of 4 centimeters, the effective focal length is 10 centimeters. Projection lenses having long focal lengths require long distances to expand the image to a given size. For a fixed projection lens, at a given distance, near or far, the image has a certain size. A 35 millimeter slide has about 3 centimeters of framed picture. By proportions, a lens with focal length 10 centimeters will expand it to 3 meters wide at 10 meters away. Zoom projection lenses are now more affordable and are adjustable in focal length so that a large image can be near or far, depending on space available. Recognizing that film prefers to curl rather than rest flat in a projector, some manufacturers provide curved field lenses that will render a properly curved transparent film image as a flat image on a flat screen.

Projection lenses can also be designed to give sharp images on curved screens. In Cinemascope projection, an interesting optical trick was played. A cylindrical lens was used to form the images on film. Wide scenes are squeezed into standard film frames, so that characters appear very thin. The wide-angle projector uses a lens designed to reverse this transformation, so that the image is spread wider horizontally than it is vertically.

The overhead projector uses a different kind of lens. Transparent material with writing on it is placed on an illuminated flat lens of large diameter. This Fresnel lens has the projection bulb at its focus and bathes the transparent object in parallel light. A conventional lens set above the object focuses and magnifies the bright object, and a mirror directs its rays to a screen on the wall. An older version, the opaque projector, illuminated book pages and other opaque material. A lens and mirror set provided a magnified but dim image on the wall. It is sometimes used by artists.

6.8 INSTRUMENTS AND GADGETS BASED ON MIRRORS

Besides telescopes, some other interesting instruments and devices make use of curved and planar mirrors. A version of the corner reflector can be made with mirrors—three flat reflecting surfaces

placed orthogonally (all at 90° angles). Large or small, it sends back toward the source any light ray striking it. This retroreflection is also obtained with small clear beads, as used in reflective highway paint. The well-known periscope uses two mirrors (or prisms) in a long tube to extend one's view. The use of curved mirrors, usually inexpensive ones, to control light sources and provide crude beams has been mentioned, as has the ability of mirrors to magnify motion. A rotating mirror will produce pulses of light in a given direction at the rate of rotation.

If one mirror gives inversion of an image (not the same as inverting it), two reflections undoes inversion, three reflections provides inversion, and so on. The change in handedness of an image is most easily traced by an extension of the right-hand rule. If the third finger is consistently pointed in the direction traveled by the actual light rays and the thumb is kept pointing up, then the image of the second finger will point right, then left, then right, and so on, after each consecutive reflection. In other words, the right hand changes to left hand upon one reflection and back to right hand after another. Elaborate mazes of mirrors have been built in Europe to exploit multiple reflections to the fullest.

Experiment 6-2

Look along two mirrors set at an angle with an object (such as a finger) behind the mirrors (see Fig. 6-14). Vary the angle, and count the number of images seen. Are there special angles where symmetrical or equal-size images are seen? Symmetry can be examined by putting a pen mark on the side of your finger tip. ■

An optical device with a long history is the "kaleidoscope." It was invented by David Brewster in 1816. As shown in Fig. 6-14, it consists of two mirrors set at an angle in a tube. Objects such as bits of colored glass are behind the mirrors so that the eye can see several reflections of them as well as the actual objects. A translucent window behind the objects will let in light without distracting scenery. The mirrors should be as good as possible, so that the difference between real object and image cannot be detected. Some images are the result of several reflections, depending on the angle. The two

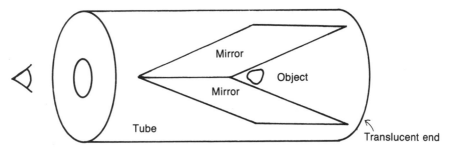

Fig. 6-14. A kaleidoscope uses two mirrors to make a number of symmetrical images of any object.

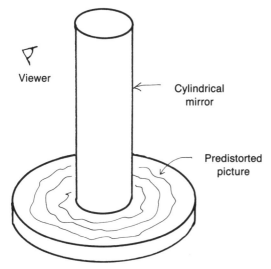

Fig. 6-15. Anamorphic art painted predistorted for proper viewing in a cylindrical mirror.

mirrors seem to fill the space with images. For angles that provide equal divisions of 360° (such as 45°, 51.43°, 60°, 72°, and 90°), eight, seven, six, five, or four identical images will be formed in a symmetrical manner. When the number of images is odd, one image must break the symmetry. The references provide more on this gadget.

Another old optical trick is the viewing of paintings in a curved mirror. For example, in Fig. 6-15 a cylindrically curved mirror (now easily made of aluminized mylar) is placed in a special location at the center of a flat, predistorted painting. The picture, perhaps a person's head with wrap around forehead and eyes, will be seen normally as an image in the mirror. The use of curved mirrors in this way is an introduction to surface mapping, whereby the contents of a square is seen as the distorted con-

tents of a distorted square or vice versa. A curved mirror is one device that can perform such a mapping. Another trick that turns distortion into normality (or vice versa) is to view a picture at an extreme slant. Some old paintings are designed to be viewed almost edgewise to see a normal scene. The references tell more about anamorphic art.

6.9 FIBER OPTICS

One of the more modern and basic discoveries in geometric optics is the construction and uses of *optic fibers*. The earliest observation of the effect was by John Tyndall in 1870 in a stream of water. In its simplest modern form an optic fiber (or fiber optic) consists of a thin rod of transparent material clad (jacketed) with a layer of another material that has a lower index of refraction (Fig. 6-16A). Light entering the fiber paraxially strikes the interior walls at steep angles, greater than the critical angle that applies for refraction from fiber to cladding. It is totally reflected and bounces down the fiber with nearly undiminished intensity. A narrow bundle of light rays entering the fiber will follow every turn and twist if the fiber does not curve with too sharp a radius. Thus the fiber conducts a set of rays from one place to another.

This principle works with an ordinary bent glass rod that has clear ends. However, the rod is not flexible like a thin glass or plastic fiber is, and any dirt or other contact with the outside of the rod changes the refraction conditions at the point of contact and destroys the reflection. Therefore the cladding is a protective layer that keeps the performance of the fiber constant and controlled. It need not be transparent but simply behave as a dielectric. By careful

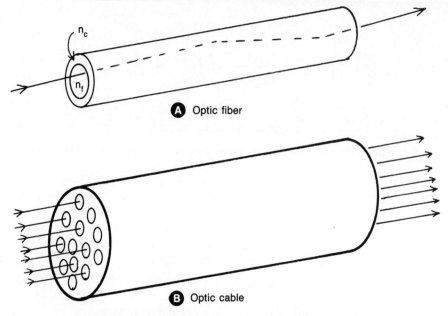

A Optic fiber

B Optic cable

Fig. 6-16. Optic fibers carry light and images with low losses.

selection of indices for fiber and cladding, rays that reflect at too small angles (measured, as usual, from the normal to the interior surface) are rejected, and only paraxial rays within a certain range will traverse the fiber. The fiber need not be cylindrical, but round fibers are easy to fabricate, and any corners such as on a square rod would trap and lose light.

Applications of optic fibers would fill several books, and many applications had been tried by 1960. Fibers are used singly or in small groups or in large sheets. An image broken down to picture elements each the size of a fiber can be conducted from one place to another as in Fig. 6-16B. Obviously the fibers must be laid carefully in order or the image will be scrambled. Sometimes intentional scrambling devices are made this way. A large bunch of short fibers formed as a faceplate or mosaic can eliminate parallax in viewing images on a cathode ray screen (oscilloscope, television, monitor, etc.). The image is formed on the inside of a vacuum tube made of thick glass, and accuracy is lost in viewing it from the outside. The front of the tube is replaced with a mosaic of thousands or millions of optical fibers, fused together and polished. If the image source is curved, as it often is, the faceplate of fibers can ren-

der a flat image. Plates of fibers can be arranged to correct lens errors, too. Fibers themselves can compress or expand images if a bunch is made of tapered fibers to form a ''light funnel.''

A small or large bunch or cable of fibers is used in the fiberscope for looking inside inaccessible areas such as stomachs. One bunch of fibers transmits viewing light to the scene and another, larger bunch brings the image back. For fine detail up to 100,000 fibers can be incorporated into a bunch less than 1 centimeter across. Traditional lenses are used to focus the object on the input end and view the tiny image on the output end.

Progress is being made on many fronts in optic fibers. The clarity is being improved so that light will travel up to 100 kilometers without needing refreshing (refreshing requires complex equipment). Compare this sort of clarity to the 6% loss of light in window glass just 3 millimeters thick. Fibers are made with variable index so that light rays rebound in smooth curves and take almost exactly the same time to travel down a fiber. Otherwise some rays become out of step with others and high-speed information flow—a major goal of optic fibers—is hindered. (Chapter 14 covers the amount of information a fiber can carry.) Some fibers are drawn to a micrometer or so, not much larger than the wavelength of light. At this size, the behavior of light changes drastically because of its wave properties. In many systems, optical information originates electrically, is converted to light for transmission; then it is reconverted to electrical signals. The information is carried in a thin, cheap fiber instead of heavy, expensive copper wire, and no electrical interference is possible while the information is in optical form.

Nature has used optic fibers in several ways, including compound insect eyes. These eyes are formed of tapered optic fibers of living transparent material that transmit an image directly to visually sensitive cells. In eyes that use lenses, the long rod and cone cells seem to function as optic fibers. In the mineral world, that there are some crystals that crystallize in the form of long, thin rods of transparent material. Root hairs on plants seem to carry light as if they were optic fibers, so that the plant knows where the soil is dark and therefore deep. Recently it has been found that polar bear hair is not white, but transparent, functioning as optic fibers to trap sunlight and channel it to warm the black skin of the apparently white bears.

Experiment 6-3

If the reader cannot find simple optic fiber in retail stores, perhaps as toys or decorative displays, the effect can be approximated by shining a bright light into the edge of a stack of glass sheets. ■

Optic fibers have many artistic applications, starting with their early arrangement as a spray of technological flowers. A hidden light source at the base of a bunch of diverging glass fibers, bent into gentle arcs under their own weight, results in bright pinpoints of light from the ends of every one.

6.10 CAMERAS AND THEIR LENSES

Sophisticated cameras use all the optical paraphernalia of the late 20th century: mirrors, lenses, prisms, stops, light sensors, coatings, light-emitting devices, even optic fibers. The widest imaginable variety of objects are converted to images on a very complex and fickle chemical sandwich.

Without photography, all science, art, commerce, and many other endeavors would advance dismally slow. There are many kinds of cameras, and only the rudiments can be described, emphasizing the optical principles. Knowledge of the principles, particularly with respect to the brightness of light, will help any amateur photographer properly expose film without the electronic aids now so popular. Unusual scenes can fool electronic sensors and computers but rarely the trained eye.

Basically the camera has a lens and film, the latter a surface sensitive to light and capable of storing an image for later development. The lens must form a real image, so the film must be just beyond the focal point. The image is inverted. By moving the lens back and forth from the film, near and far objects can be focused sharply on the film. Movement is usually accomplished through screw threads by turning the lens. There is a limit to the close focus, usually about 25 centimeters, same as the eye. Modern quality cameras imitate the eye in many ways. The field of view, normally about 40° to 50° wide, is much less than the field of view of the eye but within that range things appear much as the eye would see them. There is often a diaphragm to control brightness of the image by opening and closing an aperture.

When the aperture is wide open, it is the same size as the lens. Good lenses with normal fields of view have f-numbers of 1.4 to 2. The focal length is typically about 50 millimeters, and the lens diameter is 35 millimeters or less. The diaphragm has stops that reduce f-number in certain steps down to 32 or so. The steps or *f-stops* follow the sequence 1.4, 2, 2.8, 4, 5.6, 8, 11, 16, 22, 32. Each stop down results in a halving of the total light coming through, as can be seen by squaring this sequence of numbers. Thus the brightness received on the film is inversely proportional to the square of the f-stop (and proportional to the square of the aperture diameter).

Lenses are often interchangeable, and the variety can number over 50 different lenses for brands of cameras sold in systems. The quality must be very high and the aberrations and other image problems very low in order that a full frame of film be exposed sharply and accurately. Normal lenses are often Cooke triplets. Most lenses are flat-field lenses, designed to map each point of a flat object accurately onto flat film. Computer-assisted optical design has allowed the refinement of old configurations to compact, low-weight, multi-lens systems that are low in cost and high in quality. The lenses vary chiefly in focal length (which is related to angle of view) and in their largest aperture (which generally diminishes as focal length increases). The depth of field (range of object distances over which a sharp image is obtained without moving the lens) also decreases as focal length increases. Some forms of normal lenses are capable of very close work. The usual limit is a one-to-one translation of object to film, with the image being lifesize.

Zoom lenses were difficult to design until computer-aided methods were developed. These have an additional control that, by sliding or rotating, allows the focal length and therefore the image size to be changed without upsetting the focus. Changes in focal length from six to one are common for still camera lenses and from twenty to one for television cameras. Twenty or more individual lenses might be needed in a zoom lens to control aberrations.

A focal length shorter than normal gives a *wide-angle* view. A focal length half normal gives about 100° of view squeezed onto the same film frame. The image is clearly distorted in a fashion sometimes dramatic, sometimes unpleasant. The depth of field increases and almost everything seems in focus. An important use is for interior scenes where there is no room to step back for a normal view. Lenses that take in a wide angle of view are complex and difficult to design for high accuracy. A fisheye lens with 6-millimeter focal length compresses an extremely wide view (almost three fourths of a circle) onto a film frame meant to be viewed as covering about 40°. Such short length lenses cannot be physically put so close to the film, so the lens design gives it an effective focal length that sits far out behind the lens. In other words, the rear principal plane is far behind the lens.

Focal lengths longer than normal give *telephoto* views. For example, a 200-millimeter focal length reduces the angle of view to about 10° and magnifies the image to four times larger than with a normal lens. The depth of field becomes rather shallow and the view is quite flat with little sense of depth. Long focal length lenses are difficult to hold still without a tripod. Fast exposures are needed, yet the brightness is low. More brightness requires designing huge, heavy lenses that dominate the camera and are difficult to mount. Telephoto lenses are long, but usually not as long as their focal lengths because negative lenses at the back effectively lengthen the focal length of the objective lens group.

A mirror is a vital part of the SLR (single lens reflex) camera. It rests in front of the film, as shown in Fig. 6-17, and reflects the image to the view finder. During actual exposure of the film, the mirror swings out of the way and blocks stray light from the view finder. Then a lightweight, high-speed shutter (one of several designs) opens directly in front of the film and allows the exposure. Springs cocked when the film is wound provide the energy for these actions. Battery-powered, motorized cameras are also common now. During view finding, the mirror-reflected image undergoes another inversion in a special five-sided prism (in most designs) and enters an eyepiece to be viewed as a virtual image. The design goal of this type of camera is that the photographer see in the view finder almost exactly the image that will appear on film (except for its brightness). A frame shows the size and outline, and indicators tell the exposure being used. Larger professional cameras cannot manage to swing a larger mirror and so must resort to a view finder lens adjacent to the main lens. In the old days, the film plate was not inserted until the image had been directly inspected from the back of the camera.

Besides lens aperture, the amount of light delivered to the film depends on the focal length and the time that the shutter is open. Long focal length lenses magnify the image, and the available light is spread over the larger image, even if only part of it is seen. The loss of brightness is proportional to the square of the focal length, so that long telephoto lenses cause major problems with brightness. Lack of brightness is made up for by lengthening the time the shutter is open or opening the aperture. The possible time exposure ranges usually from 0.001

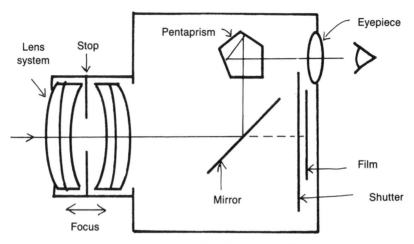

Fig. 6-17. The optical path in a typical SLR (single lens reflex) camera.

second to several seconds. Varying time or aperture has advantages and disadvantages, which must be weighed against each other in any particular situation. Camera or object motion requires very short exposures to avoid blurring. Wide-open lenses cause poor depth of field because rays have many divergent paths to travel through the lens.

One more variable is available to control final image brightness. Film is made with various sensitivities, measured with an ASA (American Standards Association) number usually between 25 and 1000. The larger the number, the shorter the needed exposure for the same image brightness (in proportion to the ASA number). Until recently, film faster than ASA 200 had definitely noticeable graininess. In two situations—long telephoto lenses used with light less strong than direct sunlight, and photography in ordinary nighttime room lighting—the available film sensitivities bring one to the limits of adequate exposure, sharpness, and small grain. When the extra requirement of large depth of field occurs for a dim indoor scene, most available films cannot meet the challenge alone.

Most cameras include at least a simple light metering system. A light-sensitive electrical sensor in conjunction with a small battery operates a small meter whose pointer is visible in the view finder. Other sensing methods are being introduced, including ones that use optic fibers. The meter does not read in the physical units of light intensity but is set for compatibility with exposure times, f-stops, and film speeds. Switching incorporated into speed and lens adjustments sets the camera to read the correct range for any exposure. The general approach is to point the camera and therefore the light meter at an object that reflects light representative of the scene of interest, adjust speed and/or f-stop to center the meter needle, than take the picture.

The fragility of cameras deserves mention. Accurate focus across the film requires very flat film held precisely in place. A focus on the film that agrees with the focus seen in the view finder is carefully set in the factory. One accidental dropping of the camera with its off-center, heavy lens can change that distance a tiny amount, ensuring blurred pictures thereafter.

7

Wave Diffraction

At this point, the possibilities of geometric or ray optics have been almost exhausted. Yet many optical phenomena remain undiscussed, such as the ability of light to sneak around obstacles. The Maxwell theory of light and other electromagnetic radiation predicts wave properties for light. Many results of the wave nature of light are almost imperceptible with ordinary light sources and optical devices but become quite obvious with the laser. How a laser works will be deferred, but idealized light sources must now be referred to that have some characteristics of laser light.

In this chapter, you will read definitions and discussions on the diffraction of waves, but you will see that diffraction and the next topic—interference—are almost inseparable. Recall that a light wave can be modeled as a moving wave front, with electric and magnetic fields alternating between maximum and zero. Wave fronts with maximum field are shown as dark lines in illustrations, and the direction of travel is always perpendicular to them. In this chapter, do not be concerned with the orientation of the fields. The wave fronts from light sources are pla-

nar for this discussion unless otherwise stated. The mathematics of diffraction is rather advanced for an introductory book, and only a few simple results will be shown mathematically.

7.1 HUYGENS'S WAVE THEORY AND DIFFRACTION

In 1690 Christian Huygens described the principle named after him—every point on a wave front serves as a source of new spherical wave fronts or wavelets. As Fig. 7-1 indicates for just a few points, this can result in quite a confusion of waves. However, the new sets of advancing wave fronts are equivalent to what the original wave front would have done if Huygens had never said anything. The backsides of the Huygens wavelets were originally ignored, but they are also present. The main reason for creating complexity in a supposedly simple situation is that the principle explains what happens when a few points on a wave front are singled out for special treatment. There is no reason to use this principle in vacuum. Rather it is applied to light traveling in materials or encountering holes or bar-

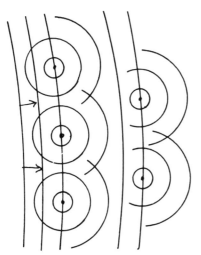

Fig. 7-1. The Huygens principle states that each wave is the source of new wavelets.

riers. The fact that materials consist of atoms with electrons, each of which reradiates light waves, fits well with Huygens' principle.

Experiment 7-1

Examine closely the shadow cast by a sharp-edged object in a strong light source. The sun or a tiny, bright lamp filament shining through a pin-hole are the best sources. Avoid a situation that produces penumbra. Examine the edge of the umbra for one or more bright and dark bands. Also try looking directly at a lamp filament through a pinhole and check for bands or fringes at the edges of the filament and the hole. ■

Huygens' principle seems to predict that waves reaching a small obstacle (Fig. 7-2A) must give rise to new spherical wavelets radiating away from points of the obstacle. The obstacle causes a change, however, so that wave fronts beyond it are not the same as the original wave fronts. Huygens' principle does not vary in its prediction of the resulting pattern as the wavelength changes, yet it is easily shown that long wavelengths (of sound, light, or any wave) travel around obstacles with less disruption than short wavelengths. Huygens' principle can also be applied to the opposite of the obstacle—a tiny hole—as shown in Fig. 7-2B. New wavelets radiate away from the hole, or perhaps from the edges of the hole.

Augustin Fresnel studied Huygens' work in the early 19th century and arrived at the necessity of including complete spherical wavelets that propagate in all directions. To make sense, this required that the wave fronts interact and add to or subtract from each other, resulting in unique but more complex

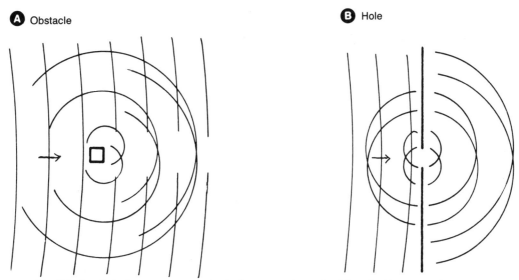

Ⓐ Obstacle

Ⓑ Hole

Fig. 7-2. An obstacle causes new spherical waves to be emitted from incident wave fronts.

waves radiating away from an obstacle and forming the patterns observed. Later the theory of light waves encountering objects was described mathematically by Gustav Kirchhoff but was found very difficult to solve in many real situations.

Different kinds of wave patterns are obtained depending on whether observation is made close to the hole or obstacle or far from it. The *near-field* or Fresnel result is especially difficult to calculate, but sometimes a description of how it works will be necessary. The simpler *far-field* or Fraunhofer result will be used wherever possible. It requires that the observation distance be much greater than the wavelength of light, a condition often easy to satisfy. The Fresnel theory is a general one good at any distance, whereas the Fraunhofer theory is an approximation where wave fronts are planar.

As a simplification of what happens when plane light waves encounter a hole, consider only pieces of wavelets generated in one direction as shown in Fig. 7-3A. One way to provide the incident plane waves is to put the focal point of a lens at a source of spherical waves. To avoid confusion, the light should also be of one color and therefore have a single wavelength. With extreme care, the separate wavelets may be called rays. An arbitrary number of points on the wave are chosen as sources of wavelets. Each point in the hole generates wavelets in step with the others. However, in the direction

shown the pieces of wavelets travel out of step and are out of phase. If combined, the net intensity would be low. There are some directions in which the wavelets are in step or in phase (for example the direction in Fig. 7-3B.) Another direction where they are in phase is straight ahead along the axis of the hole. Thus the wave intensity expected from the hole varies, depending on distance from its axis.

Figure 7-4 shows a graph (drawn sideways along the screen) of light intensity I received at position y from the axis. To avoid confusion, the hole in the diagram should be considered a long narrow slit, and everything is shown in cross section. It can be shown mathematically that there is a central peak or *maximum* of light and secondary maxima to each side, regularly spaced and growing dimmer very rapidly along the y-axis. Less than 5% of the peak intensity occurs in the first maximum on the side. The central peak is also twice as wide as the others.

This characteristic *diffraction pattern* is approximately correct whether the hole is round or a long narrow slit. The smallest dimension of the hole D has the strongest effect on the pattern. Since any number of rays from a slit can be compared for phase, the approach is to divide the width of the slit into two halves and compare a ray from the top half of the slit with one from the bottom half. Working along the slit, the same difference in path length holds for all pairs of rays (assuming the screen is

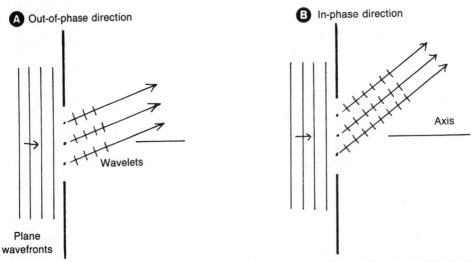

Fig. 7-3. Simplification of the wavelets generated across an aperture to find places where all are in phase and out of phase.

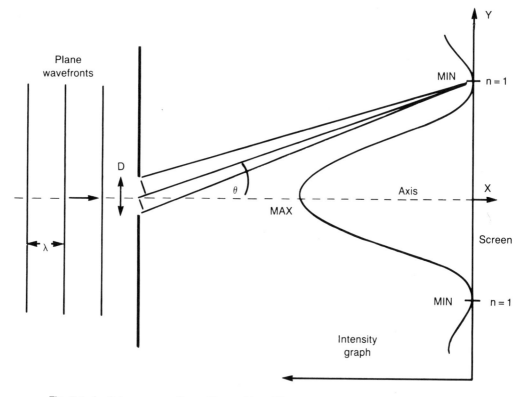

Fig. 7-4. A slit in cross section, with resulting diffraction pattern graphed at the screen.

very far away). The difference in path length depends on the direction the rays travel. Minima occur in the directions θ that satisfy the condition $D \sin \theta = n \lambda$, where n is any small integer. For a screen at x distance from the slit, $\sin \theta$ can be approximated by $\theta \cong y/x$, and the location of a minimum is given approximately by:

$$y \cong n x \lambda / D$$

The main result of this analysis is that the longer the wavelength of light used, or the smaller the slit, the more stretched out the diffraction pattern. The light and dark bands found around the central maximum in the diffraction pattern of a slit or hole (or around a small obstacle or parallel to an edge) are called diffraction *fringes*. They were first observed by Franciscus Grimaldi in the mid-17th century, who originated the term diffraction (from Latin for break-

ing up). Grimaldi did not know about waves but realized the ray model had failed. He found that the fringes were also colored, as one might expect because the various colors in sunlight would each form a set of fringes with its own spacing.

When the opening is a round or square hole, the problem is actually two-dimensional rather than one-dimensional. Wavelets are created from all parts of the edge of the hole and go in all directions toward the screen. A round hole has circular symmetry, and the pattern must show this in some way. As shown in Fig. 7-5A, light and dark rings appear around the central bright spot diffracted by a round hole. The convention for this figure only is to put dark lines where the minima are, leaving the light areas outlined for the reader to imagine as maxima. In Fig. 7-5B, two orthogonal rows of rectangular spots appear around the central peak from a square hole. The four-fold symmetry of the square hole appears in the

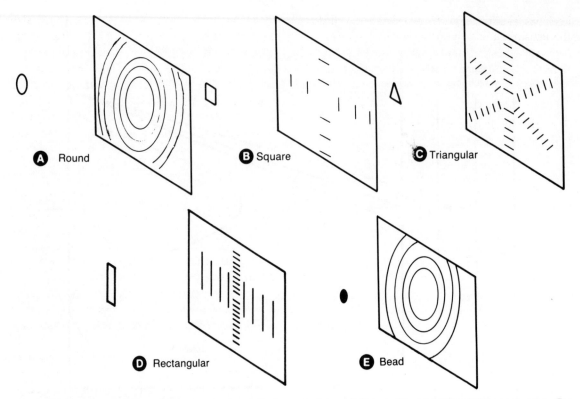

Fig. 7-5. Diffraction patterns from various shapes of aperture. Coherent light is incident from the left in each case. For this figure only, dark regions (minima) are indicated by black ink, and bright regions (maxima) are between. All but pattern "E" are from apertures in opaque sheets.

diffraction pattern. To the extent that a round hole resembles a square hole, as opposed to a slit or a shape like the letter R, the diffraction pattern will have similarities. The corners of the square hole cause a fine structure of more spots in the pattern to be explained later.

The Fraunhofer diffraction patterns of more complex holes and obstacles can sometimes be found by simple consideration of the structure of the hole or obstacle. For example, each edge of a triangular hole produces a set of fringes parallel with that edge all the way across the pattern and brightest at the center. For an equilateral triangle with three-fold symmetry, the pattern would look like a six-pointed asterisk as shown in Fig. 7-5C. For a rectangular hole with two-fold symmetry (Fig. 7-5D) the diffraction pattern has the longer fringes parallel with the longer edges and with larger spacing because the longer edges are closer together. The shorter

fringes are more narrowly spaced and are parallel to the shorter edges.

Experiment 7-2

Make a dot less than 1 millimeter in diameter with opaque ink on a sheet of clear glass or plastic. Hold it between your eye and a distant bright point of light (not the sun). Arrange for the disk to block the light source and look for circular fringes in the shadow and beyond. ■

One of the more unusual results of wave diffraction is that the diffraction pattern from a hole is similar to that from a solid round object or bead (Fig. 7-5E). Admittedly it is difficult to suspend a tiny bead in midair, but the diffraction pattern would have a central bright peak in the middle of a shadow. Waves diffracting around the bead all have the same distance to travel to the screen at the axis. The shadow itself has faint fringes within it. Outside the shadow, the usual dark and bright rings fade into the light background. The counterpart of a narrow slit is a thin wire. This complementarity of objects and their opposites occurs throughout wave optics.

7.2 PHASOR DESCRIPTION OF LIGHT WAVES

In order to understand a little more deeply how light wavelets from different parts of a hole combine to form a diffraction pattern, a phasor description of light waves must be used. As a ray of light travels, its electric field points up, then down, then up, and so on as shown in Fig. 7-6A. One period of the wave, which occupies one wavelength in distance as the wave travels, marks one complete cycle of this variation in field. The field may be thought of as an arrow or *phasor* rotating in an imaginary way, so that what is seen is the projection of the rotation as shown in Fig. 7-6B. The analogy is to the shadow of a rotating rod that falls on a screen. The locating of the phasor in time can be described in terms of what fraction of the period has elapsed or in angular measurement around a circle. By convention, phasors are started at 0° and they advance through 90°, 180°, and 270° before starting over, as shown. Figure 7-6C shows another way to depict electric phasor position as the wave advances like a ray. The length of the phasor is the amplitude of the electric field, a fixed value E.

In Fig. 7-7A, the slit (in cross section) is divided into eight parts. A ray or wavelet from the center of each part heads toward P_1 on a distant screen. They are soon out of phase. Each carries rotating phasors with amplitude E, and only the relative phases matter. What is sought is the relation of the phasors when they reach the screen and are added to obtain the net result. This is done for eight rays heading toward seven different selected points as shown on the screen, and the results of adding the phasors at each point is shown in Fig. 7-7B.

Point P_1 has been chosen so that the ray with the longest path has one additional wavelength of travel distance. Its phasor is 360° behind the ray with the shortest path as shown and so seems to be in phase. But the effect of six rays in between with in-between phases must be accounted for. Eight rays equally distributed in 360° results in a phase change of $360/8 = 45°$ between each ray and its neighbors. Phasors can be added at one instant tail-to-head as in Fig. 7-7B. The result at special point P_1 is an octagon (a coarse circle) of phasors whose sum is zero. Point P_1 must be the location of zero field and zero intensity and is therefore the location of a minimum.

For most other points on the screen, the rays from the two opposite sides of the hole will be out of phase by a certain amount. Only on the axis will all rays arrive in phase as shown for point P_0. The light there is most intense. As the position moves away from the axis, more phase difference, corre-

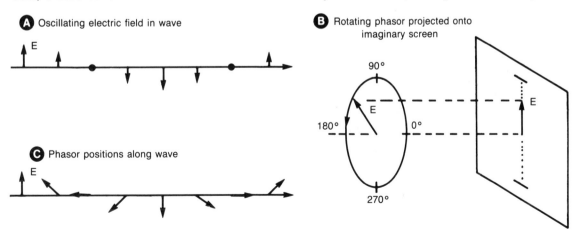

Fig. 7-6. Representation of the electric field in a light wave in terms of a rotating phasor.

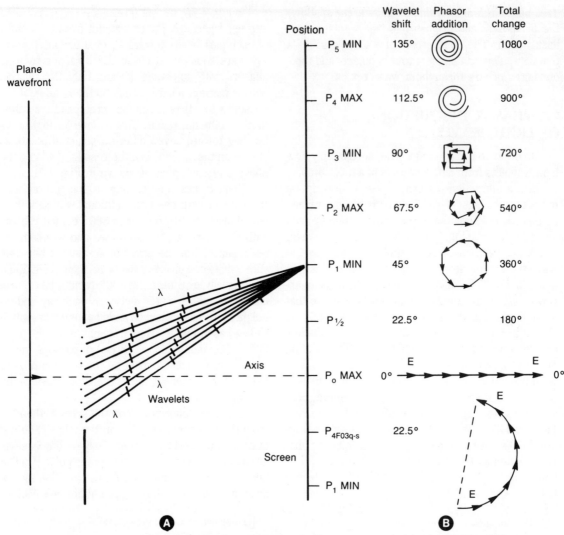

Position		Wavelet shift	Phasor addition	Total change
P₅ MIN		135°		1080°
P₄ MAX		112.5°		900°
P₃ MIN		90°		720°
P₂ MAX		67.5°		540°
P₁ MIN		45°		360°
P½		22.5°		180°
P₀ MAX		0°		0°
P₄F03q-s		22.5°		
P₁ MIN				

Fig. 7-7. Adding electric phasors for eight rays/waves leaving an aperture and arriving at the screen. Every phasor has amplitude *E*.

sponding to more path difference, occurs across the slit. The total difference advances 180° for each point, and this is to be divided among eight rays in this case. For P₂ the phase varies 540° across the slit. At P₂ the first five phasors almost close in a circle, and the next three form another half turn. The overall effect is a small net field and intensity at P₂. This is the position of the first maximum after the central one.

A second minimum occurs at P₃ where the tight spiral of phasors—shown as two interlocked

squares of phasors here—makes two turns with zero net result. For the second maximum at P₄ the spiral has 2.5 turns but is wound still more tightly, resulting in a still smaller net field and intensity. The process continues as far as needed. At intermediate points along the screen, the spiral formed by adding all the phasors has an intermediate number of turns. The net field is found by drawing a line from the head of the last phasor to the tail of the first (a chord). The spiral of phasors is called a vibration curve. One might ask what happens when the cir-

82

cle makes only one-half turn, representing the phasors landing at a point whose bottom ray is 180° behind the top ray. The result at $P_{1/2}$ is a large net field near the central maximum. This is why the central peak is broad. To obtain clearer results the slit could have been divided into more parts with more rays—there is no limit.

When all points along the screen are considered in this way, a curve of intensity shown in Fig. 7-4 is obtained. The exact height of the curve at maxima and all other points is found from the mathematical part of the analysis (see references). The minima do not fall exactly halfway between the maxima but a little farther out. The intensity of the light is found by squaring the results for the field, a process that gives the exaggerated differences between the tall central peak and the much smaller side maxima. The first maximum is about 4.5% of the central peak in intensity, the second is 1.6%, and so forth.

7.3 RESOLUTION

The analysis of Fraunhofer diffraction from a round hole is difficult mathematically but the result is simple. The angular size of the first minimum (dark ring) around the axis is given approximately by:

$$\theta \cong 1.22 \; \lambda/D$$

D is the diameter of the hole. This simple formula is important for discussing how close two holes must be in order to be distinguished. By the duality of aperture and obstruction, the same question of distinguishability might be applied to two solid beads side by side or to two bright stars close together.

George Airy did the full analysis in 1835, and the bright central spot from a hole is sometimes called the Airy disk.

Experiment 7-3

Imagine, find, or place two small, similar objects in the distance so close together that they are at the limit of being distinguished. The space between them must be less than the distance your eye can resolve. Calculate the ratio of the distance between them to the distance to you. ■

Two stars sufficiently close together appear as one star. With an excellent telescope and camera system, the two stars might be photographed as one bright spot surrounded by some complex rings of light. Such an optical system is said to be diffraction-limited, and the two objects are below the limit of resolution. To resolve them, one needs two bright, central diffraction peaks separated sufficiently that the less bright region between is discernible. In terms of intensity, graphs of two indistinguishable objects would appear as in Fig. 7-8A, forming one central peak. In Fig. 7-8B, the objects are sufficiently distinct that the central peak of one falls on top of the first minimum of the other. In this case, the two objects have an angular separation of θ as calculated above. This degree of resolution is called Rayleigh's criterion, after Lord Rayleigh, and is intended as an approximate criterion. In some cases, it might not be adequate, and in some cases careful measurement can do a little better.

Light-using microscopes had been thought to be limited to resolving distances larger than the wave-

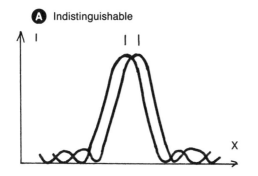
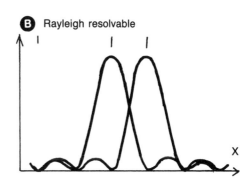

Fig. 7-8. Resolution defined in terms of diffraction peaks.

length used, due to diffraction effects. But a new technique has been recently developed called *optical stethoscopy* or *near-field microscopy*, which can do much better, distinguishing features only ⅒ of the wavelength used. The method detects points of the object through a super-fine pinhole about ⅒ the wavelength of visible light, using the central peak of diffracted light (also called the *near field*). The object is strongly illuminated and scanned point by point in 15-nanometer steps.

The resolution of the eye can be stated in terms of the ratio of the separation of two barely distinguishable objects to their distance. This is one part in several thousand. It should also equal the ratio of the wavelength of light to the aperture of the eye—in meters, approximately $5(10)^{-7}/5(10)^{-3} = (10)^{-4}$, the theoretical limit. The ratio of wavelength to aperture size can be used to estimate the angular size (in radians) of almost every sort of diffraction effect.

7.4 FRESNEL DIFFRACTION— ELEMENTARY APPLICATIONS

The general analysis of diffraction, applicable to any distance from any hole or obstacle, involves difficult mathematics. This Fresnel diffraction theory was not developed by Fresnel. A geometric method for solving such diffraction patterns was developed by Marie Cornu. Also too complex to give in detail here, the method is based on a diagram with two spirals called the Cornu spiral. It is an extension of the vibration curve used to add phasors for wavelets.

The result of using Fresnel analysis to obtain the diffraction pattern due to a single edge is somewhat unexpected. Figure 7-9 gives the intensity graph. Some light is diffracted well behind the edge into what should be the shadow region. At the edge, the field strength has risen to only half the strength for the incident light, so the intensity there is only one-fourth its full value in the light region. The intensity does not reach full value until well past the edge. Then the intensity fluctuates in a series of partial maxima and minima that grow closer together and smaller until they are lost in the general illumination.

A single edge placed in bright planar light waves lacks an essential symmetry, and the combination of many wavelets around the edge results in this complex pattern. Since edges occur everywhere in optical instruments, it would be nice to reduce the diffraction fringes which hinder image clarity. It is impossible to eliminate diffraction effects, but edges can be softened in a process called apodization. A coating at the edge of a lens that changes from transparency to opacity over a short distance will reduce diffraction there. Other tricks are possible.

A kind of lens can be designed that exploits Fresnel diffraction. It has rings of a certain width and spacing on a transparent sheet (see Fig. 7-10). If a central hole is considered and plane wave fronts strike it, the phase of all rays through the hole is about the same for any point on the axis. This is the first element or zone 1 of the Fresnel zone lens (not to be confused with a refractive lens). To bring more rays to a focus at F—a focus such that rays arrive in phase—there must be a ring-shaped region farther out called zone 3 where rays have one extra wavelength of distance to travel and therefore arrive in phase. Some variation is permitted, and the zone can have width rather than be a narrow ring. Zone 2, between zone 1 and zone 3, is made opaque so that the rays arriving approximately 180° out of phase are rejected. The same is necessary for zone 4. Rays arrive in phase from zone 5, having traveled two extra wavelengths of distance to reach F.

This alternation of dark and clear zones can continue indefinitely to create a *Fresnel zone plate*. The radius of each zone grows slowly so that the zones crowd together. The radius of the nth zone in fact is equal to the radius of the central zone times the square root of n. The area of each zone is about the same as zone 1. The light brought to point F is much brighter than the incident wave. The zone plate was

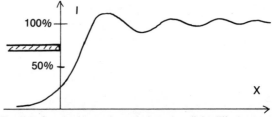

Fig. 7-9. Graph of intensity variation when light diffracts past a sharp edge.

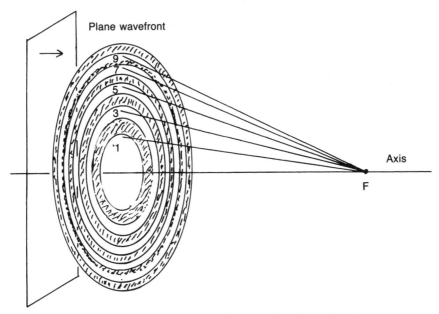

Fig. 7-10. A Fresnel zone plate focuses planar light like a lens.

invented by Lord Rayleigh in 1871. The structure is very small, and the plate is often made by printing it at large scale and reducing it photographically on transparent film. The focal length is strongly dependent on wavelength, so different colors are concentrated at quite different points. Besides the ordinary positive focal point there is a negative one representing those zones of rays that diverge from the plate but are in phase. And there are many secondary focal points closer to the plate for rays differing in path length by 2λ, 3λ, and so on.

Zone plates are capable of sharp focusing, having resolution approaching the diffraction limit. They are especially useful for focusing nonvisible light for which suitable lenses are not available. The complementary plate is just as useful. For it the odd-numbered zones are opaque, and the even-numbered zones are transparent. The central zone is opaque, and the first useful zone is the second, now transparent. The focus is in a different place, located where all rays through even-numbered zones have the same phase. In another version of the zone plate, odd-numbered zones are made thicker so that the lower speed of light in the material gives a delay equal to one-half wavelength. The result is that

all light reaching the plate is concentrated at the focus.

Fresnel zones provide sufficient explanation of a pinhole used as lens to calculate some of its properties. A pinhole behaves as if it were the central Fresnel zone number 1. The radius r_1 of the central zone or pinhole is related to the focal length of the pinhole (or a zone plate) by $f = r_1^2/\lambda$. Then the f-number of a pinhole can be calculated. Alternatively, the size of the pinhole for best image can be estimated knowing the focal length or distance from hole to screen. Typical sizes are a fraction of a millimeter.

The analysis of electromagnetic waves by Kirchhoff that led to Fresnel diffraction is not complete. Overlooked are the interaction of electric field with dielectrics and with metal edges and the polarization (orientation) of the field as it strikes the edge. Kirchhoff did improve Huygens and Fresnel theory by showing that spherical wavelets have the most intensity in front, less to the sides, and zero behind. Thus the backsides of wavelets generated everywhere on previous wavelets are of little concern.

There is an alternative to the Huygens wavelet model. The atomic model of matter to be discussed later shows that the electrons of the atoms near the edge of a hole emit spherical waves when incident light strikes them. These waves are called boundary or edge waves. The plane wave that passes through the hole simply remains that, a plane wave as wide as the hole. The diffraction pattern is the combination of the plane wave passing through the hole and the two spreading spherical waves from the two edges. The mathematical equivalence of this picture to the results of Fresnel diffraction has been shown. The electrons behind the obstacle away from the hole re-emit light back towards the source with such a phase as to cancel the forward light (if it could pass the obstacle). This sounds like a roundabout way to create a shadow—canceling forward light with reverse light—but it is a good model of how nature does it. One might wonder, however, how the front side of a building knows to cause a shadow way in back.

Experiment 7-4

Hold the sharp edge of any material in front of a bright light, just barely blocking it. Look for a glint of light from the edge, the edge wave. ■

7.5 PRACTICAL OBSERVATIONS OF DIFFRACTION

To see diffraction patterns, there is no better light source than a simple laser. Any laser beam, although small, has plane waves of light with a very narrowly defined color or wavelength. Early scientists struggled with sources such as mercury lamps. A mercury vapor lamp gives off a few distinct colors from violet to red (and ultraviolet) and therefore a variety of wavelengths, but there is sufficient variation in the wavelength in any one color to smear the results. Plane waves were often obtained by putting a bright source behind a pinhole to produce a point source. A lens then converts the spherical waves from the point source to plane waves. Other possible sources have been sodium lamps, tubes filled with hydrogen, and other ways to make selected chemical elements glow brightly at narrowly defined colors. Why elements are selected will be clearer later.

An ordinary incandescent bulb is a poor source because it emits a wide variety of wavelengths, and none of the waves are in phase. Waves come at different times from different parts of the source. As shown in Fig. 7-11, a good colored filter eliminates all but a narrow band of wavelengths, leaving light of nearly one color but with waves of various phases. A pinhole then eliminates most variation in phase due to waves coming from different parts of the source. It does not prevent phase problems arising from the spread of light after the pinhole. The light wave reaches the more distant parts of the screen at later times than the central part. The light in a central diffraction peak is adequately in phase, but the light at the edges of the central maxima and the light in the side lobes is not.

Another problem is lack of *coherence*. A light wave ideally is an infinitely long, single wave, unbroken and unchanged as it travels. But most real sources of light can emit only a short run of wave (a wave train). This wave has typically a few tens of cycles from incandescent sources or a few million cycles from atomic sources, and it passes by

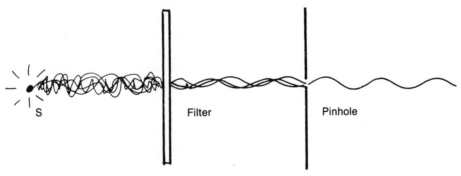

Fig. 7-11. Cleaning up white light.

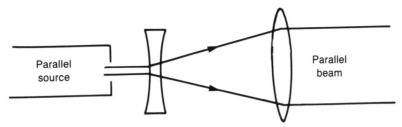

Fig. 7-12. Adjusting laser or other parallel source beam size with lenses.

quickly at the speed of light so that its coherent effects are brief. While the wave train from an atomic source might be up to a meter long, that from a laser is much longer. (Coherence is discussed more in Chapter 13.) One might wonder why any diffraction effects can be seen with ordinary hot-filament lamps.

A common laser is the helium-neon or He-Ne laser. Its light has wavelength 633 nanometers in the red. The beam is very narrow, sometimes less than 1 millimeter wide. When a wider parallel beam is needed, a diverging lens can be used as shown in Fig. 7-12 to expand it to the desired size and a converging lens to restore parallel rays.

When a laser beam is aimed at a narrow slit, say 0.1 millimeter wide, most of the light will land straight ahead on a wall (say, 10 meters away) and appear as a bright spot growing dimmer at the edges. The width of the central peak is determined by the location of the first minima on either side. Using the relation from early in this chapter, these occur at $y = nx\lambda/D = 1(10)633(10)^{-9}/(10)^{-4} = 6.33(10)^{-2}$ meters to either side. A diffraction spot about 12 centimeters wide is impressively wide. The position of the secondary peaks is difficult to calculate, but the minima that bound them are easily found to be 12.6 centimeters to either side of the axis.

If the laser is shone at a 0.1 millimeter pinhole, a large set of rings around the bright spot is formed on the distant wall. The diameter of the central Airy disk is about 1.22 times larger than the result for a slit under the same conditions, or about 15 centimeters. In a very dark room, several secondary rings will be seen.

Photographs of these phenomena are easy to make. Remove the camera lens and let the laser shine through the hole or slit and directly into the camera. Preliminary calculation will show how far away to put the camera so that the pattern of interest will fit the width of the film. Hold the shutter open for a time to be determined experimentally. Exposure time will depend on film speed, laser beam intensity, and how much the laser beam is expanded by lenses. Exposure times are typically between 0.1 and 10 seconds for film speed ASA 100. Self-developing film and camera are recommended so that improvements can be made on the spot.

An interesting application of diffraction is the determination of the orientation of tiny fibers in a composite material. As will be seen later, the diffraction pattern from many similar obstacles or holes is nearly the same as from one. If the fibers are oriented randomly, appropriate light shining through a microphotograph of them yields a circular diffraction pattern. To the extent that the fibers are aligned in one direction, the diffraction pattern will expand perpendicular to that direction and contract in the other direction.

8

Wave Interference

In the improved Huygens wave model, it was necessary to consider the effects of bunches of waves mixed together to form patterns. The main purpose of this chapter is to show how waves combine. The consequences of wave interference are shown, followed by many special applications. The distinction between diffraction and interference is seen to be mainly artificial, although the applications differ. The Huygens model is not found necessary to understand interference.

Again the type of light source needed has planar waves of one color (one wavelength). The wave fronts shown are the regions of maximum electric field (orientation unimportant). Coherence is important.

8.1 LINEAR SUPERPOSITION OF WAVES

A wave is represented mathematically with a sinusoidal variation in time and space. Most of the interest in superposition is in combining waves at one place. Then the incoming waves vary in time there. In Fig. 8-1, two spherical wave fronts are moving outward from sources S_1 and S_2. The waves have amplitudes E_1 and E_2. The wave fronts shown are taken by convention to be places where electric field is maximum and positive. Halfway between each wave front is a region where the field is maximum and negative. At the quarter-wave points the field is zero. At a fixed point, the effects of these two fields are felt moment by moment together. If an electron is there to feel the effects, it feels the resulting single varying electric field.

Where the wave fronts are shown to cross in Fig. 8-1 are regions where both fields are fully positive. At these points, there is *constructive interference*. This occurs at only a few points such as P_1, and these points move outward along the dashed line as the waves move outward. If a screen is encountered by the waves, there will be places where maximum field touches it consistently and deposits energy. There are other points such as P_2 where the maximum positive field of one wave combines with the maximum negative field of the other, giving zero. These are points of *destructive interfer-*

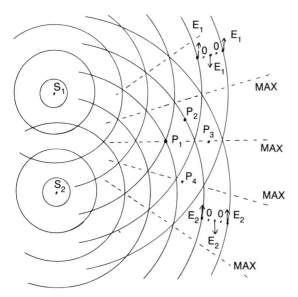

Fig. 8-1. Superposition of two sets of spherical waves (in cross section), showing points where intensity is maximum, zero, or intermediate.

ence. They will deposit no energy on a screen. At points like P_3, maximum negative fields combine constructively. These points move on the same line as maximum positive fields, and again maximum energy would reach a screen. Most points in the region occupied by interfering waves, such as P_4, are in-between cases. The fields have intermediate strengths and will add to some intermediate value. The light intensity is proportional to the square of the combined field strength at a point, and no distinction is made between negative and positive fields.

It is sufficient to track wave fronts with positive field in many cases.

The principle of *superposition* states that the effects of two or more waves can be summed at any position. The mathematics of superposition is simply algebra and trigonometry, but explicit results will not be necessary here and are left to the references. An important statement from trigonometry is that the sum of two sinusoidal functions with the same wavelength is another sinusoid with the same wavelength. The shape and wavelength (or frequency) are not changed, the amplitude and the phase are changed. The linear aspect of superposition results in addition being the relevant operation when combining waves. This may sound obvious now, but another way to combine waves is discussed later.

Another way to envision fields is with phasors. The electric field due to the waves from two sources S_1 and S_2 each varies at a position such as P_4 in Fig. 8-1. If each field has an imaginary rotation as a phasor (Fig. 8-2A), then at some instant at P_4 the fields each have a phase as shown. One can imagine the phasors steadily rotating on the axes, but it is sufficient to think of the phasor diagrams as snapshots in time. The momentary measured value for each field is the component projected onto a convenient axis (x-axis or real axis). The value is less than or equal to the amplitude. Of special interest are cases when one field is positive maximum and the other is positive or negative maximum.

Because the fields at a point can have different phases, the addition of fields has complications. Figure 8-2B shows the two phasors on the same axes

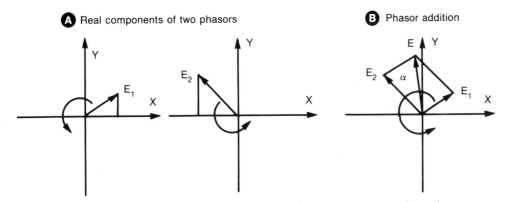

Fig. 8-2. Real (measurable) components of electric phasors, and the sum of two phasors.

so that their sum can be displayed. Geometrically, the sum is the phasor E along the diagonal of the parallelogram formed by the two individual phasors at any instant. The whole diagram rotates in time, and the phasors have a constant relation. The relative phase angle between the two fields is symbolized α. If $\alpha = 0$, the fields are in phase and the total is their simple sum. If $\alpha = 180°$, the fields subtract. They do not cancel unless the amplitudes are equal. Other relative phase angles give a total field in between. The mathematical result for the square of the field is:

$$E_2 = E_1{}^2 + E_2{}^2 + 2E_1E_2 \cos \alpha$$

The importance of this way of adding fields appears when the intensity is wanted. Light intensity is what is usually detected by instruments, the eye, or photographic film, it is a simple number without any phasor properties. It may vary in position and time. If each field has the same amplitude so that $E_1 = E_2$, then the intensity of one field is proportional to $E_1{}^2$. But the intensity of two fields together is not $2E_1{}^2$. It depends on the relative phase α. When $\alpha = 0$, the intensity is proportional to $4E_1{}^2$ or four times the intensity of one field. When the phase is $\alpha = 180°$, the intensity is zero. Other phases give values between zero and maximum intensity. A phase of $90°$ gives a total intensity twice that of one wave. Remember that these simplified results hold for waves of equal size, which is a special case. All possible intensities land on the screen when there is interference. Since no known instruments respond as rapidly as light intensity varies, they measure time-averaged intensity. Therefore at one position, the intensity seems to be constant, and an unvarying pattern is seen or photographed.

8.2 TWO-SLIT INTERFERENCE

Temporarily ignoring diffraction effects, the interference of waves from two sources is analyzed in Fig. 8-3. This is called two-slit interference (or Young's experiment from 1803) because the two sources must be in phase and have the same wavelength. There should also be similar intensities from both slits. This is best arranged by using two slits

S_1 and S_2 in front of and equidistant from a single source S. Each slit, shown in cross section, has a width D that is small but substantially larger than the wavelength of the light. The slits are a distance a apart, larger than D. Coherence is also a requirement. Continuous wave trains from the source must be sufficiently long that the same wave reaches both slits.

A screen is put a large distance away from the two slits, and the effects are observed on it by eye, photography, or some other light intensity detector. As with diffraction, a bright spot is obtained on the symmetry axis (0) and others occur at certain distances from it, wherever light waves from S_1 and S_2 arrive in phase. The path difference between the two rays must be a multiple of the wavelength λ. The locations, this time for maxima, are given by:

$$y \cong nx\lambda/a$$

where n is an integer. Only slit separation, not width, affects the pattern. For $n = 0$, the maximum is on the axis. This interference pattern is graphed with intensity sideways on the screen. The positions 1, 2, and so on are maxima. The peaks continue regularly and indefinitely to the side in the ideal case. The minima are exactly halfway between the maxima. The minima are completely dark only if the two slits provided equal intensities.

The slit width enters into interference in the following way. If only one slit is open, and the hole is not much larger than the wavelength, a broad central diffraction peak is obtained as in Fig. 8-4, centered on the open slit. If instead the other slit is open, a similar peak is obtained centered on it. If both slits are open, the interference pattern occurs where the diffraction pattern permits. Even more unexpected than this result is that the diffraction pattern for two identical holes is a single peak the same size as for one hole and centered on the symmetry axis between the holes. It is not ''spread'' by the two holes.

The diffraction pattern, as graphed in Fig. 8-4, is an envelope for, or modulates, the interference pattern of the two slits. The brightest interference

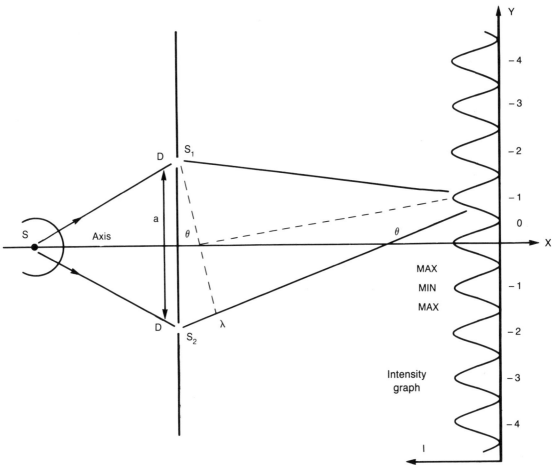

Fig. 8-3. Two-slit interference (Young's experiment) in cross section, with path difference shown for first off-axis maximum and intensity pattern on screen.

fringes are visible where the diffraction peak would have been. To obtain many interference maxima from two slits, the slits must be far apart (resulting in close spacing) and their size must be small, giving a wide diffraction peak. When the slit width is one-fourth the separation as shown ($a/D = 4$), seven maxima occur within the central diffraction peak. The interference maxima continue to the sides, much diminished, wherever side lobes of the diffraction pattern permit. Viewing two-slit interference with a laser is easily carried out and shows this modulated pattern rather than a long series of identical maxima.

Experiment 8-1

Punch two pinholes less than 0.5 millimeters each in foil 1 to 2 millimeters apart. Use a distant bright light source and put the foil very close to the eye. Is there a pattern of fringes visible? Which way is it oriented? ∎

Seeing interference with ordinary light sources is best done by using a distant source through a colored filter. A pinhole is a way to convert a close bright source into a tiny dim source. Small pinholes will let only a few coherent wave trains through at one time. These may be up to a meter long. A col-

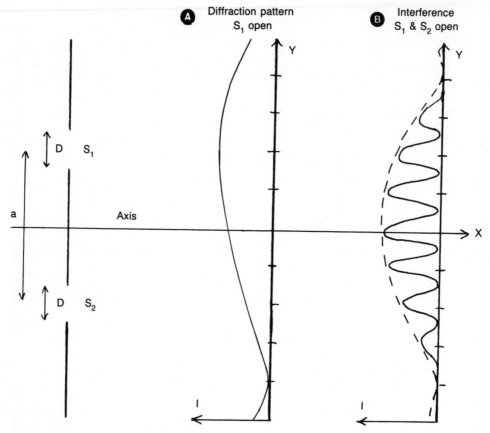

Fig. 8-4. The diffraction of one slit modulates (determines the envelope of) the interference pattern when two slits are open. Ratio of slit width to spacing is four.

ored filter restricts the range of possible wavelengths so that many differently colored maxima are not superimposed.

Diffraction through two round holes also gives an interference pattern. Figure 8-5 illustrates the astounding result that the diffraction pattern for two holes is the same as that for one located on the axis. The separation of the two holes results in bright and dark straight fringes across the diffraction pattern. Holes that are spaced four times farther apart than their width result in a pattern with closely spaced fringes, seven fitting across the central Airy disk. The more holes, the brighter the pattern, and the more interference fringes modulated by the diffraction pattern. The diffraction pattern of a sieve of identical round holes is a single Airy disk with rings

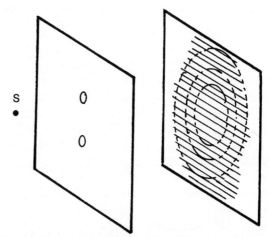

Fig. 8-5. Representation of the diffraction pattern from two round holes separated four times the hole size.

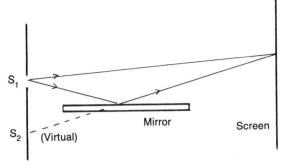

Fig. 8-6. Interference from one source using grazing reflection from a mirror (Lloyd's mirror experiment).

as if there were just one hole centered on the axis of the source.

The equivalent of two-slit interference can be produced with mirrors. In the arrangement called Lloyd's mirror (Fig. 8-6), the source is viewed by two paths, one direct and one reflected at a glancing angle in a mirror (plain flat glass will do). The reflection reverses the phase 180°, but otherwise the analysis and the fringes are the same. Two mirrors with a slight angle between them (say 178°) can be arranged so that one source is seen in the mirrors as two. This Fresnel's mirror also shows interference fringes. Interference also occurs from dust specks on an ordinary back-silvered mirror, resulting in a pattern of rings. The front and back surfaces of the mirror participate to form this Whewell-Quetelet pattern.

8.3 MULTIPLE-SLIT INTERFERENCE

Any number of slits can be combined for interference. Very interesting optics occurs when there is a long array of sources, usually obtained with a long series of identically spaced slits and one planar light source. Eight are shown in Fig. 8-7. The slit-to-slit spacing a need not be much larger than the slit width D. As slits are added, the principal maxima that appear with two slits grow more and more narrow. They continue to appear in the same positions, as determined by $y = n\lambda x/a$. Other secondary maxima from the interference among many slits appear between them and become smaller and smaller with ever finer spacing.

If there are N slits, then there are $N-2$ secondary maxima between the principal maxima. For eight slits, six secondary maxima appear as shown. As slits and secondary maxima are added, increments of width are taken away from the principal maxima, but the principal ones always appear to be twice the width of the secondary ones. All maxima are observable only within the envelope of the diffraction peak and side lobes for the particular slit width. As the number of slits increases, the principal maxima are separated by more dark space. They assume the appearance of lines and are called *spectral lines* corresponding to the wavelength used.

Multiple-slit interference can be analyzed further by considering the field of the wave from each slit as a phasor for any chosen point on the screen. For example, eight rays are shown heading for position 2. As usual, the principal maxima are formed from rays which are in phase. The analysis is very similar to that for diffraction from one wide slit but the resulting pattern is different. Figure 8-7 shows how the eight phasors from eight slits are added for several positions on the screen. At position 0 on the symmetry axis, the rays from all slits are approximately in phase because the screen is very far away, the pattern very wide, and the maxima narrow. Here the amplitude and intensity are largest and form a principal maximum (which has $8^2 = 64$ times the intensity from one slit). Position 2 is the first minimum, when eight consecutive phasors add to zero in a circle. Each is advanced an angle of 45° over its neighbor.

At position 1, the phase differences must be half those at position 2, so the ray from each slit is 22.5° out of phase with its neighbors. At position 3, the phase differences have advanced to 67.5°. A little less than six phasors close a circle, and two more go almost halfway around again. The net phasor is rather short, not much longer than a single one. Geometry is needed to find the exact length. A secondary maximum is obtained at position 3. When the fields are squared to yield average intensities, this secondary is less than one-tenth the principal. The analysis can be continued for any points where the relative phases are conveniently known. At position

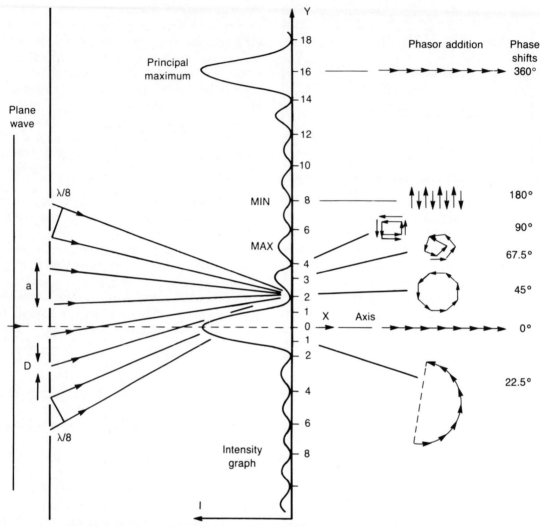

Fig. 8-7. Interference pattern and addition of selected sets of phasors for eight slits.

eight, each ray is 180° out of phase with its neighbors. At position 16 (a principal maximum) the phase difference between any 2 slits is 360° (corresponding to one wavelength path difference).

Ways to make devices that function as multiple slits include using fine parallel wires (the complement of slits), reflections between two parallel surfaces, and cutting fine grooves in a transparent or opaque sheet. For some applications, it is desirable to have each source in an array differ by a definite phase that can be varied. No optical version exists as yet, but radar and radio telescopes are good examples of

what may come. As the phase between each tiny transmitting antenna in a phased-array radar is varied, the location of maxima is scanned across the sky. Interference patterns work in reverse, too. If a row of detectors, such as radio telescopes, all look in the same direction at planar waves from a distant source, the ability of the system to distinguish directions improves as the number of detectors increases. Similarly, the direction in which the telescopes look can be changed, not by rotating the telescopes physically, but by varying the phase differences inserted in the signals from each.

8.4 GRATINGS AND SPECTROSCOPES

When the slits are sufficiently numerous, it does not matter if the separation is just a little larger than the slit width. The practical limit is reached with a *diffraction grating* (or just grating). A grating is made as in Fig. 8-8A by ruling many tiny grooves on a plate of glass. The unruled parts remain transparent and function as slits. The glass is rough in the grooves and traps or scatters the light there. Approximately 1,000 lines per millimeter are possible, and gratings can be over 10 centimeters wide. Joseph Fraunhofer seems to have been the first to make and study gratings. The grating might just as well be called an interference grating as the distinction between diffraction and interference is moot for this optical device.

When planar light is shone through the grating, the grating is used in transmission, and light of a particular wavelength is deflected in a few unique directions. If the glass has a thin aluminum coating that is scratched off when the grating is ruled, then it is used as a reflection grating (Fig. 8-8B). A particular wavelength, instead of reflecting in the direction a mirror would take it, is diffracted to a few unique directions. In analysis the same angle of diffraction θ occurs in both cases and is given by the grating equation:

$$\sin \theta = n\lambda/a$$

For accuracy, the *sin* function should not be approximated by just the angle θ. Usually the case $n = 1$ is the only one of interest. Light of the same wavelength is diffracted to larger angles determined by values of $n = 2, 3$, and higher, and is dimmer. The *order* of the diffraction is determined by the value of n.

The grating equation shows readily that light of shorter wavelength is diffracted to smaller angles. The colors of light are widely separated, and a grating is an excellent way to display a spectrum. A bright spot of undiffracted light also occurs on the axis of the source. This source light is not only wasted but hinders seeing the dim colored spectrum. Modern precision gratings introduce an additional parameter or way to affect the interference pattern into their design. Figure 8-8C shows a blazed grating with a zigzag reflecting surface. Incident light is shone normal to the grating as a whole, and the angle or blaze of the surface is made so that reflection occurs in approximately the same direction as diffraction for wavelengths of interest. Thus most of the source light is "encouraged" to appear as spectral lines rather than as an undiffracted bright spot.

The importance of the number of lines in a grating enters when the resolving power or ability to discriminate between nearby wavelengths is considered. This chromatic resolving power is defined as the ratio of the wavelength to the increment in wave-

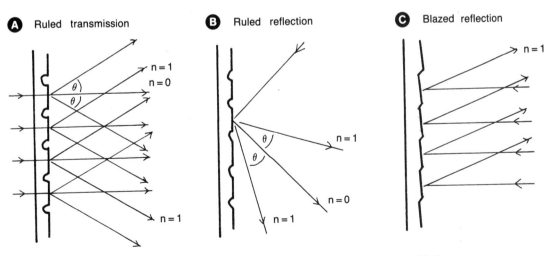

Fig. 8-8. Three kinds of diffraction gratings (much enlarged profiles).

length that can be distinguished. Simple analysis shows that this power is equal to the number of lines ruled on a grating, provided that all the lines serve as slits for the incident light used. The number of lines can be as high as 100,000 so that wavelengths can be measured to an accuracy of one part in 10^5. Each of the thousands of slits (lines) in a grating contributes literally to the accuracy.

A simple setup for using a grating is shown in Fig. 8-9. The optional lenses (or collimating telescopes) expand the point source to illuminate the entire grating, then collect the parallel diffracted rays to form images of the slit as spectral lines. To give an idea of the sizes involved in using a grating, the first order angular location of red light of 700-nanometer wavelength is computed here. A grating with a typical 6,000 lines per centimeter has a line separation of $1.67(10)^{-6}$ meters. Then sin $\theta = 700(10)^{-9}/1.67(10)^{-6} = 0.419$, or $\theta = 24.8°$. Violet will appear at half this angle because its wavelength is half as much. As before, diffraction is very wavelength-dependent.

For amateur use, high quality replica gratings can be obtained at extremely low cost. These are thin transparent plastic films with the grooves molded in. They are made from a master ruled grating used as a mold for plastic. Good gratings may have ghosts—dim unwanted and spurious maxima. Explaining these constitutes a good introduction to the types of error that can occur in optical systems. If dirt, defects, and irregular ruling occur randomly on the grating, then there is no systematic pattern to cause diffraction to maxima in particular directions. Instead, principal maxima are broadened by random scattering of light. If the error in ruling is systematic, perhaps from the ruling engine making the same slip in the same place each time it rules a line, then there is basis for a new diffraction pattern, and maxima will be produced in the wrong places.

8.5 THIN-FILM INTERFERENCE

There are many other ways to produce interference of light—that is, to set up conditions whereby the same light or two sources of related light are superposed in the same place. In *thin-film interference*, reflections from front and back of a thin layer of transparent material give a pattern of

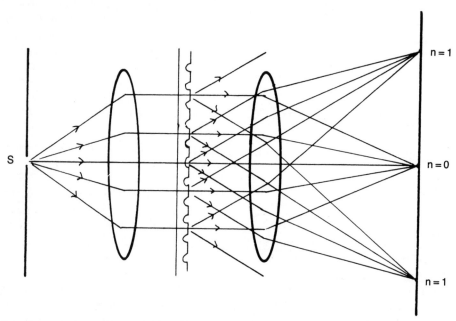

Fig. 8-9. Using a transmission grating to obtain a first-order spectral line from a slit source. Lenses to spread the source light over the grating, then collect parallel rays, are optional.

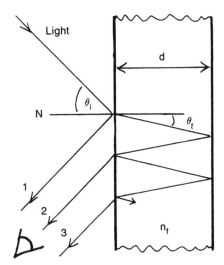

Fig. 8-10. Thin-film interference from multiple reflections. Parallel fringes are obtained when the film is uniform.

fringes. As shown in Fig. 8-10, incident light at some angle is reflected toward the observer by two or more paths. Path 2 is longer than path 1 by the distance and extra time of travel in the film. At larger angles of incidence (measured from the normal), the extra path will be longer. To complicate matters, the angle of travel inside the film is smaller in this case due to refraction.

The light in paths 1, 2, 3, and so on, interferes after it leaves the surface and produces bright and dark fringes—maxima and minima. For smooth uniform films, they are parallel bands oriented perpendicular to the plane of incidence. For maxima the extra distance traveled must be an integral multiple of the wavelength. The result of analysis is a relation between angle of incidence and various parameters:

$$\cos \theta_t = m\lambda/4$$

The angle θ_t is in the film as shown. It can be converted to an external angle θ_i by Snell's law and the index of refraction of the film. The wavelength λ has the value that pertains within the film. Maxima are obtained when the integer m is odd, and minima when it is even. The relative phase shift of 180° between externally and internally reflected rays has been used to obtain this result. Since the interfer-ence relation is different for different wavelengths, different colors from the incident light will be sorted out and will appear as colored fringes. A common example is sunlight shining on a thin oil layer on water. Because this surface is not flat or still, the fringes appear as iridescent colored swirls. The colors of the fringes can sometimes be used to estimate film thickness.

The destructive interference that produces dark fringes in multiple reflections from a film needs some explanation. The first and second reflections for a minimum are out of phase, but the first is brighter than the second, and they do not cancel. However, the second and all higher order reflected rays are in phase and taken together are just strong enough to cancel the first ray.

Thin films can be of water, air, or coatings on glass. Films can be of constant thickness or wedge-shaped. Two sheets of glass propped apart at one end by a piece of paper will show interference in the air layer, and the fringes will spread farther as the line of observation approaches the thinnest part of the layer. Such fringes are called Fizeau fringes. Ring-shaped fringes called Newton's rings are observed when a gently convex lens is placed on very flat glass. This was once the way to measure accuracy in optics shops. Optically flat glass is glass ground and polished to much less than a wavelength in flatness and is necessary to observe orderly interference fringes and well as to measure defects. An ordinary glass sheet placed on an optical flat will show irregular interference fringes telling where the surface of the ordinary glass is high and low.

For normal reflection from a thin film, it is possible to select thickness and index so that all reflection is canceled at a particular wavelength. This constitutes an *antireflection coating*, which is much needed on lenses in complex systems. The thickness, together with the index, is chosen to give a quarter wavelength path (or any odd multiple of a quarter wavelength). As shown in Fig. 8-11, incident light entering the coating and reflecting from the back surface travels an extra half wavelength and returns out of phase with the light reflected from the front surface. Another requirement for complete cancellation is that the amount of light reflected from

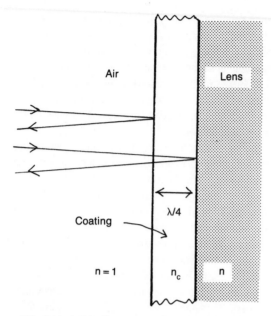

Air

Lens

Coating

$\lambda/4$

$n = 1$

n_c

n

Fig. 8-11. A thin film as an antireflection coating.

front and back be the same. This holds only if the ratio of indices is the same for both reflections. If $1/n_c = n_c/n$, then the coating must have an index n_c equal to the square root of the index n for the glass. This coating will be antireflective from both sides.

A common problem in multiple lens systems is *flare* or the reflection of bright light among lens surfaces. Coatings as described below will prevent most reflection. Such coatings are usually calcium or magnesium fluoride and will appear reddish violet. The color is due to their being selected to cancel colors centered around yellow. The antireflection works adequately for light incident at modest angles and cuts reflection from 4% to about 1% per surface. It follows that the light transmitted increases by this amount. In a ten-lens system, the effect is cumulative for each lens. Much more light passes through the system. Nevertheless, multiple coatings have been found necessary to reduce further the flare in elaborate zoom lenses.

Coatings with many layers are *interference filters* (or Fabry-Perot filters) as well as superior at antireflection. These are tuned in their design to transmit a selected range of the spectrum and re-

flect the rest, or vice versa. Usually the stack of films has each layer a quarter wave thick, and layers are of alternately high and low index material. Eight layers designed to enhance reflection perform better than the best front-silvered mirror. Of interest is the cold mirror, which separates heat from light, letting heat radiation through while reflecting light. The opposite design is also used—reflecting heat while passing light. Unfortunately fabrication is too expensive for large-scale control of solar energy. Interference filters are also found on peacock feathers and the backs of beetles as irridescent colors that change with viewing angle. Analysis of the performance of multiple layers of thin films is best done with matrices if there is computer assistance. For each film, there is a transfer function, and these are multiplied together to obtain the result for many layers.

8.6 INTERFEROMETERS

Any optical system that provides interference between light waves from two or more places gives a means of making precise measurements, employing the wavelength of light as a ruler or of measuring wavelength itself. Already mentioned has been the examination of Fizeau fringes for irregularities in a supposedly flat surface.

The Fabry-Perot interferometer (after Charles Fabry and Alfred Perot in 1897) uses multiple reflections in an air gap between two parallel partially silvered flat surfaces (Fig. 8-12). The air gap can be varied. The partial mirrors allow some light to leak out from each reflection so that multiple parallel rays of similar intensity are produced. The source is intended to be broad—that is, made of many independent sources as from a glowing gas tube. Light of one wavelength entering the gap at a particular angle undergoes multiple interference and ends as sharply defined rings on the screen centered around the lens axis. The higher the reflectivity, the sharper the fringes.

This interferometer can produce much sharper fringes, and therefore spectral lines from a source than diffraction gratings or other interferometers and needs slit. The intensity contour of a fringe is called an Airy function. The Fabry-Perot interferometer

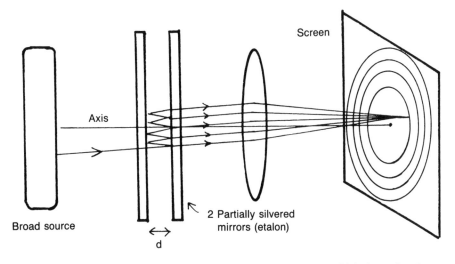

Axis

Broad source

2 Partially silvered
mirrors (etalon)

d

Fig. 8-12. Schematic arrangement of the components of a Fabry-Perot multiple beam interferometer.

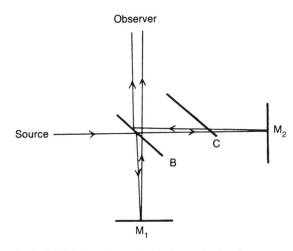

Observer

Source

C

M₂

B

M₁

Fig. 8-13. Michelson two-path interferometer (angles exaggerated). Fringes observed are rings or parallel bands.

has about 10 times better resolution than a diffraction grating. Two nearby wavelengths are distinguished as two sets of circular fringes on the screen, with the radii of the fringes slightly different for each set. The Fabry-Perot interferometer does not show clear useful patterns with white light. It is intended to resolve spectral lines already selected with a diffraction grating or other means. When the air gap between the partial mirrors is fixed, the unit is called an *etalon*.

The Michelson interferometer (from Albert Michelson in 1882) compares light rays sent by two different paths. In Fig. 8-13, light from source S is divided approximately in half by the half-silvered mirror or beam splitter B. Part is reflected from the backside of B toward mirror M_1, and part is transmitted toward mirror M_2. Both reflected rays return to be superposed and viewed by the eye. The ray from M_2 passes through a compensating thickness of glass C so that both paths travel the same distance in glass. Then this ray is reflected from the silvered surface to reach the eye. The two light paths are long, but if they differ by some fraction of a wavelength, this will be detectable. The eye normally sees a set of circular light and dark fringes because rays passing through the system at various angle have various path lengths that give constructive or destructive interference. The fringe pattern is similar to that from the Fabry-Perot interferometer except the fringes are much broader. Again, white light does not produce fringes; only light restricted in wavelength will produce fringes.

The Michelson interferometer is used to measure details of spectral lines because slightly different wavelengths will produce fringes of different radii. It also can measure length in terms of wavelength. A very small motion of one mirror can be measured by counting the movement of fringes. As

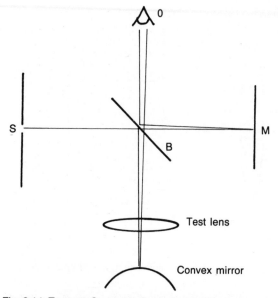

Fig. 8-14. Twyman-Green interferometer used to test a converging lens for spherical accuracy against a spherical convex mirror (angles exaggerated).

a mirror advances a half wavelength, one fringe grows from or disappears into the center spot. Accuracy approaches 1% of the wavelength used. A mirror can also be tilted slightly so that the fringes are nearly straight lines. This interferometric technique has been of great value in testing whether light travels in an ether (it does not) and for relating the length of a standard meter to a wavelength of light. The laser is often the light source of choice, although interferometers do not absolutely require coherent planar light. Interferometers are very sensitive to vibration and usually must be mounted on heavy tables isolated from external shaking and noise. Interferometers with one very long path are in use to measure movement of the Earth at a fault zone, among other applications. Length is limited by variations in the air, so evacuated tubes are often used on long runs.

A modification of the Michelson interferometer is the Twyman-Green interferometer (from 1916) for testing lenses for imperfections. As shown in Fig. 8-14, a test converging lens is put in one path of the interferometer. Plane waves passing through it are converted to waves showing distortion due to the test lens. A perfect lens would produce spherical waves from plane waves. A convex spherical mirror is located so that spherical waves from the lens are reflected as spherical waves. They will be converted to plane waves after returning through the lens. Distortions are doubled. If the test object is supposedly flat and the mirror really is, then the object is tested for flatness.

Other uses of interferometers are measuring the index of refraction of materials, the thickness of films, and the coherence length of light. The lamellar grating, consisting of two sets of mirror strips that are interleaved and can be moved, forms an interferometer that needs no beam splitter but uses a concave mirror.

8.7 MOIRÉ PATTERNS AND OPTICAL BEATS

A convenient and graphic tool for illustrating wave interference is the use of Moiré patterns (literally, watered or wavelike patterns). These result from superposing regular grids or arrays of straight or curved lines. In many cases the interference of wave fronts is simulated in cross section. As Fig. 8-15A shows, two sets of parallel lines, not necessarily with the same spacing, form fringes from plane waves. Two circular rulings show similar interference (Fig. 8-15B). Circular rulings in the manner of a Fresnel zone plate evoke ghosts of themselves from parallel lines (Fig. 8-15C). Other grids of lines with varying spacing are useful or at least make interesting patterns.

Moiré patterns are capable of large magnification and therefore have optical uses. Any difference between two supposedly identical sets of lines or dots is readily apparent when they are printed on transparent material and brought near alignment. There are ways to use the patterns to test optical components such as lenses. A microscope can be made with two sets of ruled lines and no lenses. Two-dimensional patterns and three-dimensional surfaces can be analyzed by projecting rulings or grids on them, somewhat in analogy to laser-based methods covered later.

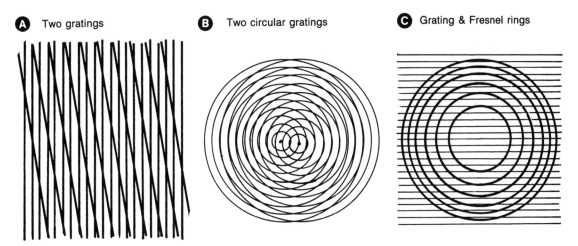

A Two gratings **B** Two circular gratings **C** Grating & Fresnel rings

Fig. 8-15. Moiré patterns formed from parallel lines, concentric circles, and Fresnel rings.

Experiment 8-2

Moiré grids can be copied onto transparent film with xerographic or thermal copiers. See the references for ideas; then experiment with simulated wave interference using superposition of various patterns. Half the fun is in moving the patterns. Try making and using colored ones, too. With some perseverance, Moiré transparencies can be found for sale. You may have noticed everyday Moiré effects from loose-weave cloth or screen wire. ∎

Optical *beats* result when two waves with different frequencies interfere in time. The conditions and theory are covered more in Chapter 15. Beats have been undetectable with visible light until recently because of the need for very steady and pure waves. Theoretically two extremely narrow-band waves which differ by, say, 10 Hz will beat at that rate. The result would be a 10 Hz pulsation in brightness.

9

Interaction of Light and Matter

You have read about the elementary aspects of the interaction of light and matter in regard to laws of reflection and refraction, as well as diffraction phenomena. If there were no matter present, whether in the form of the air around all equipment, in the form of smooth shapes of glass with assorted coatings, or as many natural surfaces, light waves once produced would continue forever unchanged. In the near vacuum of space, which is bathed in strong sunlight, the "background" appears black in almost all directions. If light did not interact with matter, you would not detect its presence with photography, detectors, or the eye. Although some colorful phenomena are introduced here, details of color are left to Chapter 17.

This intermediate discussion of light and matter recognizes that matter is made of atoms with electrons but treats those components classically—that is, as simple mechanical objects. Quantized aspects of atoms and light will be discussed in Chapters 11 and 12. For the present, an atom is treated as a building block of nature whose negatively

charged electrons are all that are of interest. It is important to note, however, that most of the weight in an atom resides not in its electrons but in the nucleus at the center. Lightweight electrons seem to fill most space in matter and help to bond atoms together to form liquids, crystals, metals, and other types of materials. Often two or three atoms bond strongly to each other to form a molecule, which then may be a building block of larger forms of matter. Review the brief discussion of how vibrating electrons emit light in Chapter 3.

9.1 LIGHT WAVES IN MATTER

Does light somehow shine straight through transparent materials, perhaps passing among electrons without disturbing them much? The evidence that light does not pass casually among electrons is that its speed is substantially lower in matter, with a definite value for each material as measured by an index of refraction. To reduce the speed of light, it seems that strong interaction with the electrons in the material is required. Recall that the index for air

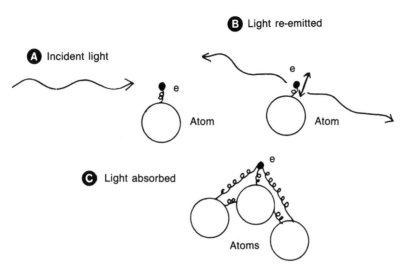

Fig. 9-1. Light wave emitted and absorbed by an electron bound to an atom. The binding is modeled with springs here.

is very near 1 (about 1.0003). As a gas, air is mostly empty space populated with molecules rushing about at high speed. Only about one part in a thousand of a volume is occupied by air molecules, and the index is different from the vacuum value by a similar amount. The empty space in air is a reason the speed of light is affected so little in air.

On the large scale, light waves striking matter have a choice of being reflected, transmitted through, or absorbed (converted to heat). On the small scale, a light wave striking an electron in matter vibrates it at the frequency of the light (Fig. 9-1). The force on the electron comes from the varying electric field in the light wave. The magnetic field that is part of light is much smaller in effect and can be neglected in virtually every case. The actual movement of the electron is very small, usually less than one millionth the size of the atom. Depending on the relation of the electron to nearby atoms (binding is modeled with springs here), it will either re-emit the same frequency as it vibrates (Fig. 9-1B) or it will transfer the energy to nearby atoms and electrons (Fig. 9-1C). The lightweight electrons are easy to vibrate, but most of the weight is in the nuclei of the atoms, and they move too slowly to respond to light waves.

Bound electrons in atoms like to vibrate, but the vibration frequencies they prefer—as required by the particular atom—are usually quite different from the light. If the natural or resonant frequency is higher than the frequency of the light, the electron will vibrate along with the light. If the resonant frequency is lower than that of the light, the electron will oppose the vibrating force imposed on it. In this case, and when the light frequency is very near resonance, the electron absorbs light wave energy by passing it around to nearby electrons and to the atom.

Because a light wave affects many electrons simultaneously, there is also the possibility of large-scale effects from the reactions of many electrons. Vibrating as a group, the electrons can scatter re-emitted light in any or all directions. Light scattered back is seen as reflected. Sometimes it is scattered in an orderly way and therefore constitutes *specular* reflection like a mirror (Fig. 9-2A). Often it is scattered in a disorganized way and is seen as *diffuse* reflection (Fig. 9-2B). There is a preference for scattered light to reflect in the direction it would take according to the law of reflection. The surfaces of mirrors are difficult to see because you see only the orderly reflected images. The surfaces of diffusely reflecting materials are easy to see because we can see their color and surface texture.

As noted earlier, light striking any surface at a large (grazing) angle tends to be scattered specularly. When a rough surface is viewed at a grazing angle, its roughness is seen compactly. The many widespread bumps appear close together and only

103

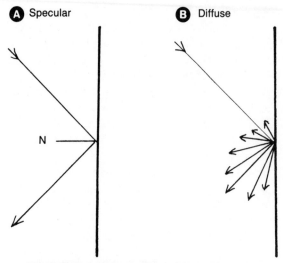

Fig. 9-2. Two kinds of reflection from surfaces.

their tips are seen. On the average, the bumps function as a mirror regardless of the particular atomic composition.

Transparent surfaces by definition are almost colorless. They pass most of the light that enters them and preserve the order of the light rays. A lens works because the rays passing through it are not absorbed or mixed up but permitted to keep the order of the image they represent. The resonant frequencies of transparent solids and liquids are not in the visible spectrum, or the effects would be seen as color. Most resonances are in the ultraviolet. Water is resonant in the infrared, too. Colored glass has a range of resonances at visible frequencies and therefore it has an *absorption band* of frequencies. When light is absorbed by a transparent or opaque material, the light frequency must have been near one of the resonances of the electrons in the material. The effect appears in bulk form as a rise in the index of refraction for that wavelength or frequency. In transparent materials, the index increases with frequency as the resonances in the ultraviolet are approached. Absorption is stronger as resonance is approached, so that blue and violet light are slowed more than green and red.

Diamond has one of the highest indices (2.42), and light is drastically slowed in diamond. This is related to the tight bonding of electrons in diamond—such bonding manifesting itself as hardness and the difficulty of dislodging a carbon atom. The appearance of diamond, at least when cut to certain shapes, is a result of the many more angles at which total internal reflection occurs and the greater dispersion.

Glass, long the most popular optical material, is not a crystal and not a liquid but something amorphous in between—a frozen liquid that never formed a crystal structure. It is somewhat unstable, so that local damage readily propagates to destroy larger parts of the glass. Glass making was discovered about 7,000 years ago by accident, and glass working began about 2,300 years ago. Now, more than a 100,000 different formulas for glass have been tried. But glasses are very complex, and understanding them is a major challenge.

9.2 MORE ON LIGHT WAVE PROPAGATION IN MATTER

Often the atoms and electrons in a material have an orderly arrangement. In Fig. 9-3, a very simplified crystal is shown with atoms as an orderly set of dots in two dimensions. A plane light wave, consisting of vibrating electric field, attempts to pass through this crystal. Each atom has electrons that vibrate in step and emit spherical waves like Huy-

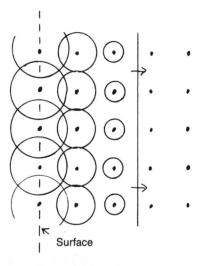

Fig. 9-3. Plane waves traveling inside matter consist of superposition of wavelets from each atom.

gens wavelets. The spherical waves spread in a complex but orderly maze of wave fronts. The leading wave front is an envelope for the spherical waves catching up with it. The orderliness of the crystal is a perfect setup for both destructive and constructive interference.

To this point, nothing has been said that would reduce the speed of light. Incident light and spherical waves propagate at c. Their combination or superposition, however, has a lower speed. The spherical waves are not in phase with the plane wave but rather lag behind it. For various reasons including dissipative losses, the electrons respond slower than the incoming wave fronts. The spherical waves also vibrate electrons and further delay accumulates. The combined effect is a plane wave propagating more slowly through the material. The original plane wave is canceled out. Moreover the final wave moves forward in the direction of the original (with the direction altered as required by matching wave fronts at any boundary). All spherical waves moving backward and to the sides cancel each other by destructive interference.

Reflection occurs in the top layers of atoms or molecules at the surface of a material (Fig. 9-4). The reflected light is the backward spherical waves radiated by the electrons in the top layers of the material. These combine to form a departing wave front that obeys the law of reflection. The wavelets from

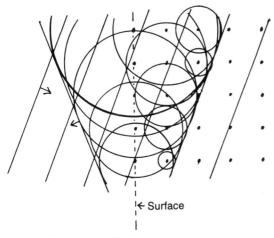

← Surface

Fig. 9-4. Plane waves reflecting from a surface are composed of wavelets in phase.

deeper atoms begin to interfere destructively and may not reach the surface. Or the atoms are so absorptive that no light penetrates that deep. Reflection from a solid or liquid does not involve scattering by molecules acting independently. Large numbers of molecules interact with each other and reradiate light in an orderly fashion. The surface material that is one-half wavelength deep—nearly 1,000 atoms thick—is called the Fresnel zone and contributes to most optical effects. (To avoid confusion with the zone plate, this term will be avoided here.)

When light arrives at or near Brewster's angle (the polarizing angle), the electrons near the surface vibrate as usual. Those vibrating in the plane of incidence find they are vibrating in the same direction as any reflected ray is supposed to depart. Therefore they are unable to emit a suitable electric field that can become part of a light wave departing in that direction. When the angle of reflection is near Brewster's angle, then the electrons can manage only a small amount of vibration in the proper direction (perpendicular to the direction of the parting wave).

Another interaction between light and electrons at the surface of a material is an action-reaction effect. The field of the light wave pulls the electrons back and forth a bit. But they are anchored to the atoms of the material and resistive to moving. This results in a reactive pull on the field, and any reflected light wave departs with a phase shift. The phase shift is 180°—complete reversal of the field. There is one exception for waves arriving at angles larger than Brewster's angle and in the plane of incidence. Then the electrons produce a small amount of field upside down, so to speak, and the phase shift is changed to zero (reflection in phase).

What happens when a wave travels in matter and reaches the surface at such an angle that it emerges parallel with the surface? The light has no choice but to propagate as a surface or evanescent wave. It cannot be internally reflected because the incident angle has not quite reached that value. The surface wave does not spread far into the second material (or air) but stays near the surface, eventually being all absorbed in interactions with the surface atoms.

Unusual radiation occurs when objects—atomic particles—are sent through a transparent material faster than the speed of light in the material. This Cerenkov radiation is blue-green and is emitted like a shock wave—an optical sonic boom. Electrons passed by the high-speed charged particle are impacted by the passing electric field and then resonate at their natural frequencies, emitting light.

9.3 HOW MATERIALS APPEAR IN TRANSMISSION AND REFLECTION

Generally the electrons in materials re-emit light more efficiently in the blue and violet than in the red and orange. However, the top layers that determine reflection depend on wavelength. Blue light penetrates less deeply to begin with, so less electrons participate. And the area over which atoms interact constructively to reflect light depends on wavelength, too. Less area participates in reflecting blue. Therefore all colors are treated about the same and many substances can appear white in reflection. More exactly, the reflection has the color of the light source. A transparent material such as glass reflects a small amount of light without color preference. Deep in the glass, the color is slightly blue-green from the effect of electrons reradiating.

Although the main resonances of water molecules are outside the visible range, there is mild absorption of red light. Water molecules are polar in that they have a separation of charge, making them very susceptible to electric fields. Infrared light is strongly absorbed. Figure 9-5 shows the absorption curve (or, conversely, the transmission curve) for water in the visible and near-visible range. Half the red light is absorbed in a few meters of water. In deep clear water, the remaining light after red absorption tends toward blue-green. Ocean floor creatures often are red, although in their environment without red light they appear black.

When materials that are ordinarily transparent are broken up into small pieces as powders or grains, these appear white in the aggregate. They do not contain enough ordered atoms to reflect light specularly. Therefore they reflect diffusely with a slight preference toward specular. None of the many reflections and refractions of light rays that wander into the pile of pieces has a preference for any visible wavelength, so the material remains white. This description also pertains to materials made of fibers, such as paper. The molecules of cellulose and other fibers in paper and textiles usually have no resonances in the visible so there is no coloration

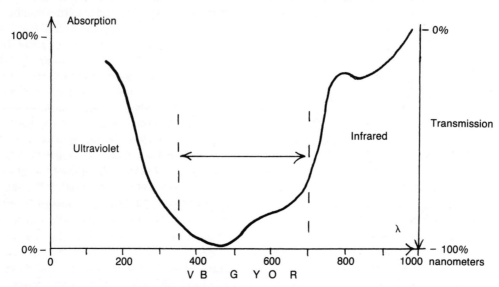

Fig. 9-5. Light absorption curve for water.

until inks and dyes are added. Molecules that give color are surprisingly rare.

At the opposite extreme is the attempt to make a black object. By ''black'' is usually meant that all visible light is absorbed. Carbon and certain carbon-based molecules are among the few excellent light absorbers. Space scientists have now found this effect almost three billion kilometers away on the moons of Uranus. Most black objects still reflect about ten percent of the light. Flat black is little improvement over glossy; it just scatters more diffusely. Good black surfaces such as velvet have deep texture on a small scale to trap scattered light.

Study of the way light is reflected by metals is of interest for explaining how they have their particular metallic colors and for understanding why mirrors are coated with metal. In metals many electrons are not identified with particular atoms but roam freely as a volatile sea of electrons. They are free to conduct current, heat, and other properties. Light waves will affect them strongly, and they are free to vibrate more vigorously. Ultraviolet light is sufficiently energetic to eject electrons from the metal, as will be seen later as a quantum effect.

The electrons in metals automatically vibrate out of phase with light because an electric field pushes the negative electrons in the direction opposite the field direction. Therefore the electrons reflect light almost perfectly, and no light penetrates the metal deeper than about a wavelength. The reflection is broad-band so that shiny metal is a good reflector of infrared, including long-wave heat radiation, as well as some ultraviolet. If the metal surface is smooth with irregularities smaller than the wavelength, the reflection is orderly (specular), and the surface is a true mirror. Some metals reflect without preference in wavelength and appear silvery (for example, silver, aluminum, steel). Transition elements such as copper and gold have other electrons in their atoms which resonate in blue light. These absorb green, blue, and violet to some extent, leaving red, orange, and yellow to be reflected.

A mirror can be partially *silvered*—that is, coated with a thin layer of aluminum or silver on the front or back. When the thickness is just less than the wavelength of light, about half of the light will penetrate the silvering and the other half will be reflected. This forms a beam splitter so necessary for interferometers and other optical instruments. Glass that reflects over a broad band is made by including small amounts of metal in it or coating it with a thin film of metal. If the film is thin enough, it does not reflect visible light well but will reflect long wave infrared, thus functioning as a selective heat reflector. A little energy is absorbed by metals instead of reflected, or metals would not feel hot in the sun. Certain dark oxide coatings on metal can change this. The metal absorbs over a broad band and emits very little infrared although the metal becomes very hot. This arrangement is called a selective surface and is good for collecting solar energy.

9.4 ATMOSPHERIC SCATTERING

A vibrating electron re-emits light with different efficiencies at different frequencies. The intensity of light emitted is proportional to the fourth power of the frequency of vibration (or inversely proportional to the fourth power of the wavelength). That is a very strong effect. The scattering of light by this process is called Rayleigh scattering, after Lord Rayleigh (a 19th century physicist). Visible light passing through gas such as air causes each loose electron on each molecule to reradiate light at the various colors, but much more strongly in blue and violet light. Because visible wavelengths are much larger than molecules (by a thousand times), Rayleigh scattering results in spherical waves from each molecule regardless of its shape. Our sky appears blue because the blue light in sunlight is most strongly scattered in all directions including toward us (other colors pass through with much less effect). The sky does not appear violet because sunlight is more deficient in violet than blue to start with. The skies of other planets appear different colors—e.g., yellow-orange on Venus because little blue light penetrates the thick air; blue-green on Uranus because gaseous hydrocarbons absorb the red and yellow.

At the other end of the spectrum, the sky shows much red light when the sun is low. In this case we are looking toward the sun and seeing the least scattered light. The green and blue have been scattered the most by the extra thick portion of air we look through, leaving red and orange predominant. By the same reasoning, distant snowy mountains, a source of white light, should appear tinted orange yellow at any time of day. No discussion of the appearance of the setting sun can go without mentioning the green flash. The last light to be refracted over the horizon after the sun has truly set is blue and green. Blue is strongly scattered, so a brief green flash remains.

The effect of Rayleigh scattering is weak because air is not dense with molecules, but without it the sky would be as dark as space. Nevertheless, about 10% of the light energy coming from the sky is sunlight scattered from air molecules. Light can travel very far (over 200 kilometers) in the atmosphere before its energy is scattered away. Line of sight is hindered more by the curvature of the Earth and the thinness of the atmospheric layer than by the way air molecules scatter light.

An important optical component of the atmosphere is water. Whether vapor, cloud droplets, raindrops, or ice crystals, it strongly affects visibility and also causes many remarkable displays in the sky. Water vapor is a gas and affects light the way air molecules do; it is simply another rare molecule. Water droplets are a semiorderly substance with molecules packed closely so that they can interact. Scattered light is unchanged in color, so the droplets appear white from sunlight. Hence clouds appear white (and shades of gray in shadow).

Experiment 9-1

Observe the color of the smoke from a thick pile of burning leaves. Compare the color at the edge and center of the smoke plume. Is there any black smoke? ■

Matter so finely divided as to be smaller than light wavelengths but larger than molecules will appear bluish from Rayleigh scattering. Thin, fine smoke particles will appear orangish if the light source is behind the smoke because the blue light has been scattered away by the smoke. Thick smoke, however fine, must appear white (or gray) because all colors are eventually scattered uniformly in it. But matter is usually not so finely divided so wavelengths cannot be preferentially scattered. Light waves scattered by larger particles in *Mie scattering* are scattered—really, diffracted—mostly in the forward direction, whereas molecules scatter light equally in all direction. Particle sizes can be so small that the material is not directly visible, yet the particles are still much larger than the wavelengths of visible light. They will appear white in sunlight. If the sky does not appear deep blue (except at the horizon), particulate matter is scattering light. ''White'' skies are not due to clouds but often are the result of particles from human activities. Recently the general distribution of sizes of particles in our atmosphere has been found to occur in two ranges—particles from 10 to 1000 nm and those from 1000 nm to 0.1 mm. Particles around 1000 nm, comparable to near-infrared wavelengths, are rare.

More on the scattering of light from small particles, including those suspended in liquids (for example, milk), is left to the next chapter because polarization effects are significant. Black smoke is due to absorption, not scattering of light. Carbon particles (soot) preferentially absorb shorter wavelengths, too, so that light through black smoke may appear reddish or brownish.

Mirages in the atmosphere have received extensive discussion in the literature (see references). The variety of possible imagery is large and fascinating. Another name for the desert mirage (where warm, less dense air is near the ground) is *fata morgana*. The results include sky as water, floating mountains, and upside-down ships. Images appear lower than the objects. Opposite in effect is the artic mirage or *hillingar*, where cool dense air is near the ground or ocean. Images appear higher than the objects and so can be seen for abnormally long distances. Mirages would be more extreme on a planet like Venus with its superdense atmosphere.

9.5 RAINBOWS, GLORIES, AND OTHER ATMOSPHERIC DISPLAYS

Water droplets and ice particles cause some fascinating light shows in nature for which explanation

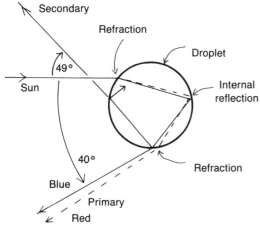

Fig. 9-6. Sunlight refracting and reflecting in a water droplet, emerging at the most likely angle. Large numbers of droplets give the rainbow because different colors emerge in slightly different directions from a droplet.

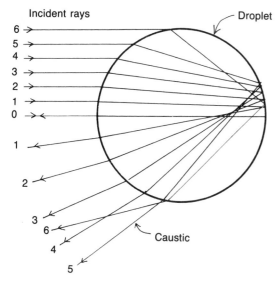

Fig. 9-7. Caustic resulting when rays enter a droplet at various positions and angles. Ray 5 emerges near the rainbow angle of 42°.

is now possible. Water droplets small enough to be suspended in air or to fall slowly as rain are nearly spherical. As shown in Fig. 9-6, sunlight is refracted, reflected, then refracted when it strikes a droplet, resulting in a separation of colors. Each droplet produces spots of colors, but a sky full of droplets gives an orderly pattern. Measured with the sun

directly behind the observer, the spectrum from violet to red is spread in an arc called the *rainbow*. The outer size of the arc is about 42°. The distance and width of the arc cannot be measured in meters because it is a virtual image of the sun dispersed into colors and arranged across the sky. A rainbow can have no physical end, unfortunately.

Sunlight can reflect more than once inside the water droplets before emerging dispersed. Many possibilities have been observed in experiments with single droplets. The secondary rainbow is due to one additional reflection in each droplet. This time blue emerges at a larger angle than red, and the whole 'bow is located outside the primary one, 50° in angular radius. René Descartes provided the first explanation of rainbows up to this point. Inside the primary rainbow are a faint series of pink and green bands called *supernumerary arcs*. They seem to be due to a more complex effect than refraction. Another effect not widely realized is the fact that the sky is darker than normal between primary and secondary 'bows. This dark arc is called Alexander's dark band.

The explanation of Alexander's dark band lies in a consideration of what angles a ray can have when it emerges from a droplet. Rays can enter the droplet in a range of angles from head-on to glancing (Fig. 9-7). Their paths are not ordered in a simple way through the droplet but cross over. The emerging rays tend to bunch up as if focused, forming a *caustic* (Latin for burning). There is a maximum angle (the rainbow angle) at which rays emerge, and a dark area (Alexander's) occurs next to the caustic where no rays go. Much work has been done on calculating the intensity of the various rays in order to predict the intensity of various parts of the rainbow. Its brightness is not uniform from red to violet. Droplet size is also found to affect the intensity of each color.

Experiment 9-2

Examine the concentration of rays when sun or bright light passes through any curved, transparent material such as a lens, marble, or bottle with a complex shape. Why does it help to fill the bottle with water? ∎

The supernumerary pink and green arcs depend on the size of the droplets, unlike the 'bows. Large drops give narrow closely spaced arcs. This sounds like an interference effect, and Young showed this early in the 19th century. Rays reflected from different places in the droplet have different path lengths and can interfere when seen superposed. When white light encounters an opportunity for interference, only certain colors are seen to interfere in certain directions. The pink and green arcs appear for the same reason pink and green bands are often seen in thin-film interference. Wave theory has been applied to the rainbow problem as a whole to find what happens when planar light waves strike a transparent sphere, and work has proceeded on finding the answers and comparing with observation. The rainbow is not confined to visible light. Infrared counterparts occur along with the visible displays.

Another spectral display seen in the direction opposite the sun is the *glory*. Again light is refracted in water droplets, but this time the effect is from those rays that strike the droplet at a grazing angle and are refracted inside as shown in Fig. 9-8. After an internal reflection, the ray is refracted back but not quite straight back. The ray can also do fourteen reflections inside and return similarly. The light is refracted back to the outside of the droplet at an angle tangential to the droplet. It follows the surface as an evanescent wave until it is reinforced in returning the way it came. The rays are returned from all around the edge of the droplet, so it looks like a bright ring.

Each color has a different small distance to travel on the surface of the droplet in order to achieve constructive interference, so each color is reinforced in a slightly different direction. The edges of droplets, seen as tiny glowing colored circles, produce the glory. Each acts as a diffraction aperture, whose resulting pattern is a series of rings like Fresnel zones. Since many colors are involved, the rings are colored in a complex order rather than being light and dark. The same pattern is obtained for any number of tiny bright rings spread randomly in the sky. Since the image is virtual, the pattern is centered on the observer's line of sight away from the sun and can only be measured angularly.

The glory is best seen from the air, looking into misty clouds or fog away from the sun, exactly where the shadow of the airplane falls. The smaller central rings are the most brightly colored, and the pattern fades out at large angles. Observing from an airplane or mountains are about the only ways

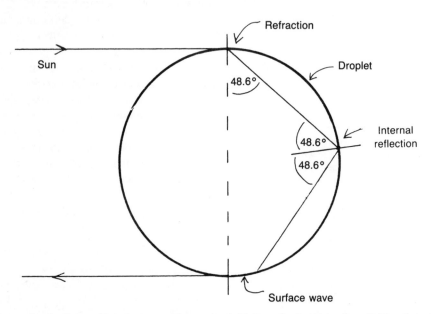

Fig. 9-8. Internal reflection and interference makes a droplet into a glowing ring of sunlight and produces the glory.

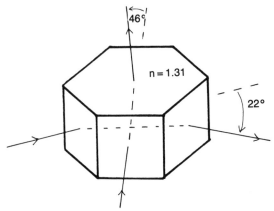

Fig. 9-9. Hexagonal ice crystals high in the atmosphere refract and reflect light to produce halos, arcs, and other phenomena.

one can get between the sun and misty air. The wave theory of the glory begins similarly to the rainbow and was solved (in principal) by Gustav Mie in 1908. Fast computers are needed to obtain practical results. Experiments with single droplets have shown that laser light striking one edge returns straight back from the other edge, having followed the surface of the droplet for about 14°.

Small ice crystals high in the atmosphere around the sun refract sunlight toward the observer. The refraction disperses light, so that rings, arcs, and spots of colored light can be seen in many positions around the sun. While some sources call all these phenomena *halos*, this term will be reserved for circular patterns. But all are due to the ice crystals that constitute cirrus clouds. There are also pillars of unrefracted light. Ice crystals have just a few common forms, all based on the hexagon. They are either hexagonal rods, rods with one end brought to a point (like a six-sided pencil sharpened with six knife strokes), or hexagonal disks (like bathroom floor tiles). The index of refraction is 1.31, slightly less than the index of water. As shown in Fig. 9-9, light can refract through ice crystals in two basic ways—entering one face and leaving through an opposite one, or entering a face and leaving through an end. When a ray enters and leaves a crystal or any other prism in a symmetrical manner, its direction is deviated the least. The effect of minimum deviation produces a caustic of rays. That is, there is a range

of preferred entry directions that result in similar exit directions so that the rays are concentrated in certain directions.

Experiment 9-3

Draw a hexagon on paper and use ruler and protractor to draw and test various rays entering the hexagon, refracting twice, and leaving. Comparing outgoing ray with incoming, what seems to be the minimum angle of deviation for all rays you can test? (You will need a calculator, Snell's law, and the index 1.31 for ice to find where refracted rays go.) ■

When the crystals are randomly oriented, the result is a ring or halo around the sun (Fig. 9-10) Light refracting through the 60° faces is deviated 22° from the direction of the sun and forms a halo with angular diameter 44°. Light refracting through a face and end is deviated 46° and forms a halo 92° across. Both halos have dispersion with red on the inside and blue outside. The color separation is not as complete as with a rainbow. The inside edge is sharp because of a caustic; the outside edge is fuzzy. Recently, long-disputed and rare observations of "Scheiner's halo" with an angular radius 47.5°, unobtainable from hexagonal crystals, have been confirmed by laboratory identification of octahedral (8-sided) ice crystals. These can produce the requisite angle of refraction.

Depending on the size of the crystals, they are buffeted by air currents to assume random orientations, or they fall with a particular orientation—the long axis down. When the crystals are oriented downward, there is no halo but instead four or more bright spots to the side of the sun. These parhelia or sundogs are images of the sun formed a few degrees outside where the halos would be. Parhelia and halos are the most common atmospheric optical displays seen from ice crystals.

Also notable is a pillar of unrefracted light extending vertically from the sun. The light is reflected by the ends of falling oriented crystals. The crystals wobble during fall so that the reflection is smeared vertically. If one looks in the direction opposite the sun a spot called the anthelion is present. To the sides are two spots called paranthelia. A faint circle of light, the parhelic circle, can some-

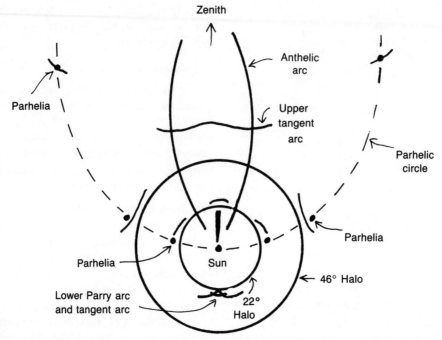

Fig. 9-10. All the more common and some of the uncommon observable halos, arcs, and spots around the sun.

times be seen around the sky passing through these spots and parallel with the horizon (constant altitude). It is not refracted light but light reflected from vertical sides of crystals all around the sky.

Seeing more elusive arcs and spots around the sun depends on its height above the horizon. An *arc* here is a short curved bright line in the sky. There are at least nine kinds of arcs (some shown in Fig. 9-10), and most are due to falling ice crystals that are spinning. Some can be seen only from airplanes because they are below the horizon otherwise. Research continues on both observing and explaining

unusual arcs. Halos, arcs, and spots tell much about the state of the weather high in the sky and have some valid relations to weather prediction. Halos are also commonly seen with moonlight.

A "corona" (not to be confused with the intrinsic corona of the sun) is due to diffraction from water droplets near the sun, giving rings with red outside and blue inside. The colors are not very distinct, but the middle is yellowish. Corona is not easy to see because of the sun's brightness, but it is quite apparent around the moon. Corona size depends inversely on water drop size.

10

Polarization

The fact that the electric field in a light wave has orientation has appeared during the discussion of many sorts of phenomena. This topic must now be discussed in detail. Light observed from scattering, from rainbows, from reflection, and from many other interactions of light and matter has a property you cannot see but simple instruments and animals can. This property of polarization is a broad topic with many ramifications, both theoretical and practical. With caution you may use the wave and ray descriptions of light interchangeably. Rays do not have polarization, but they conveniently describe the direction of travel. Any ray must be understood to have a light wave oscillating perpendicular to it. This is the last chapter in a series in classical optics based on rays and electromagnetic waves and ignoring quantum effects.

10.1 LINEAR AND CIRCULAR POLARIZATION

Polarization is defined as a specific orientation of the electric field in a light wave. A number of light

waves added together may or may not have an overall polarization, too. A jumble of light waves from an incandescent source is said to be unpolarized because the orientations of the electric field are random in both space and time. The average electric field at any time is likely to be zero, although the average intensity is not. A plane wave in a single wave train, must be polarized as shown in Fig. 10-1A. In this figure and all later ones, the magnetic field of the light wave is not shown, but whatever polarization the electric field has is true for the magnetic field also. The right-hand rule relates the polarization of the two.

Polarization can also be depicted as in the right side of Fig. 10-1A by looking down the light wave in the direction of travel and noting the direction of the field E. The field is shown for the positive half cycle and is understood to reverse every half cycle. There is a plane of polarization in which E vibrates, shown dotted. Since E is at an angle to the axes, components of the electric field can be taken along the axes. Two such orthogonal polarizations add to

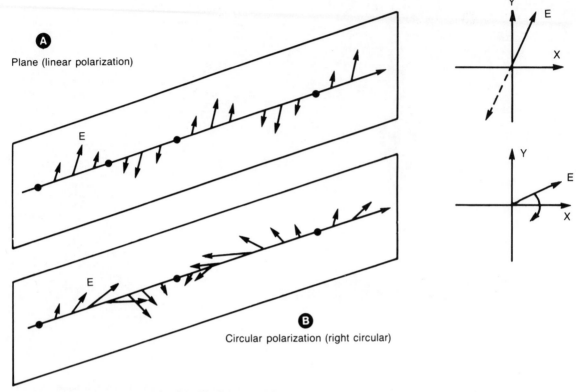

A Plane (linear polarization)

B Circular polarization (right circular)

Fig. 10-1. A plane or linear polarized light wave and a rotating or circular polarized light wave.

zation if the components are in phase. If the plane remains fixed in time, the light is *linear polarized*.

Sometimes the light wave is *circular polarized*. In this case, the plane of polarization rotates over time, and the field points somewhat as shown in Fig. 10-1B. The period of rotation is the same as the period of the light wave itself. Circular polarization comes in right- and left-handed types. The right-hand rule can be used to determine the type. If the thumb is pointed in the direction the wave travels, then the curl of the closed fingers points in the direction of rotation for right-handed polarization. Left-handed polarization is opposite, and it matches the left hand.

By adding or superposing two linear, polarized waves with equal amplitudes and certain orientations and phases, circular polarized light can be created (Fig. 10-2A). The two linear polarized waves should be 90° out of phase and oriented at right angles. In the figure, the two fields are tracked for about a quarter cycle, showing that the net field turns. If the direction of travel is into the plane of the diagram, this arrangement whereby the horizontal field reaches a maximum 90° after the vertical field produces right circular polarization. The amplitude of the rotating field does not vary in this case. When the amplitudes are different for the two out-of-phase linear waves, their combination is elliptically polarized (fatter in one direction than the other).

In the converse to constructing circular polarization, adding left and right circular polarized light of the same amplitude together produces linear polarization. In Fig. 10-2B the fields of two oppositely circular polarized waves are shown at some instant. The net field is also shown. Later both original fields have rotated equal amounts and the net field diminishes to zero, then reverses. It remains in a plane and thus represents linear polarization. The uses of these elaborate descriptions of polarization will be more apparent in applications.

114

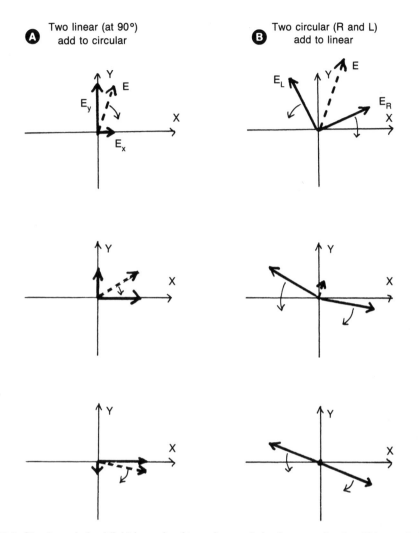

Fig. 10-2. Circular polarized light is made of two plane polarized waves vibrating 90° out of phase, and plane polarized light is equivalent to two opposite circular polarized waves. Snapshots of the superposition of the fields is shown for about one-fourth cycle.

Experiment 10-1

Obtain two polarizing sunglasses and look through two lenses in series. (Or remove the lenses from the frame of one pair. They will be useful throughout this chapter.) Rotate one lens with respect to the other. If the light through both does not grow dimmer and brighter, one or both pairs are not polarized. A greenish color is no longer a sufficient indicator of polarized sunglasses. How many times does the light go dim as you rotate one lens through 360°? ■

10.2 POLARIZING MATERIALS

Various devices that convert mixed random unpolarized light to polarized light are *polarizers*. One kind is made of a complex material named Polaroid. Besides this common, thin, greenish plastic, there are reflective and scattering methods of polarizing. Nature has provided some crystals that prefer light oriented in particular directions. *Linear polarizers* convert random light to linearly polarized light. These devices have a direction associated with them and might be imagined to have a line painted on them

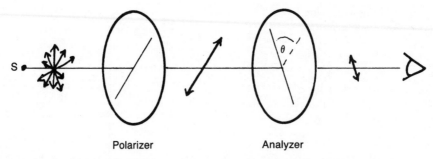

Fig. 10-3. Polarizing random light, then analyzing it for polarization direction with a second polarizer at a different angle.

that tells the polarization direction or *optic axis*. Other polarizations are more or less strongly suppressed by polarizers, usually by absorption. Random light is reduced to half intensity by a polarizer.

To determine if a device is a linear polarizer, the resulting light can be tested with a second (known) polarizer as shown in Fig. 10-3. The second polarizer is called an analyzer. The polarizer converts random light to polarization in one plane. The analyzer allows only the component of light to pass that agrees with its optic axis. When both polarizers are aligned parallel, all the polarized light passes through. When the analyzer is turned to a right angle with respect to the polarizer, no light (ideally) can pass through both. Thus rotating one polarizer in front of another shows a sequence of light fading to dark and increasing again, twice during one full rotation. The law regarding these intensity changes is the law of Malus, after Etienne Malus in 1809. Mathematically it states:

$$I = I_0 \cos^2\theta$$

where I is the resultant intensity, I_0 is the original intensity, and θ is the angle between the two polarizer directions. The *cos* function is one for parallel polarizers (zero angle) and zero for right angles.

Passing light through a grid of fine wires illustrates how one might polarize light. The wires conduct current and allow electrons pushed by the electric field parallel with the wires to move. The electrons consume the energy in light waves for any field pointing in such a direction. The light waves are left with electric field, if any, perpendicular to the wire direction. This polarization direction perpendicular to the slots between the wires may seem counter intuitive. The experiment with wires has actually been done with infrared light, but the wire size and spacing must be about the size of a wavelength.

Natural crystals such as tourmaline have electrons that prefer to vibrate in one direction. Some materials provide one-dimensional electron conduction paths on a very small scale and work as polarizers. Materials that have a property that depends on one of the three independent directions possible in the material are called *anisotropic*, in contrast to *isotropic* materials, which show no directional preference. The disadvantage of crystalline polarizers is that they absorb various colors differently and do not preserve white light.

The first polarizing substance, Polaroid (technically, herapathite), was invented in 1928 by Edwin Land. The material in use today is an improved version (H-sheet) developed in 1938. It imitates a wire grid, using iodine atoms attached to long polymer chains to make them conducting. The material is formed mechanically as a thin sheet with this alignment built in. It is almost impartial to wavelength but polarizes blue light less efficiently so that dark appears bluish. A perfect polarizer is called HN-46 because 46% of the incident intensity is passed and 46% absorbed. The other 8% is reflected by front and back surfaces.

10.3 POLARIZATION BY REFLECTION

Another way to polarize light is by reflection, discovered by Malus in 1808. A consequence of analyzing wave reflection is the discovery that there is a certain angle of incidence that results in polarized reflected light (recall Chapter 3 discussion).

This occurs because only one polarization direction is reflected (the case when E field is normal to the plane of incidence). The light polarized parallel with the plane of incidence is totally transmitted past the boundary and cannot be reflected because there is no field component available to form a reflected wave.

Experiment 10-2

To find out why polarizing sunglasses were invented, examine the reflected glare of sunlight from metal or glass (on a car) while rotating the sunglasses through 360°. What might it mean if you find no effect? ■

Near the polarizing or Brewster's angle of incidence, reflected light is partially polarized, having the normal component emphasized due to loss of most of the parallel component (Fig. 10-4). Thus sunlight shining on a horizontal shiny surface is partially polarized horizontally. Looking at the surface through polarizing sunglasses with the optic axis vertical would eliminate much of the reflection. Intensity is reduced to the degree that the angles of

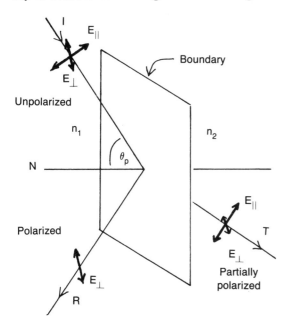

Fig. 10-4. Using Brewster's angle to polarize light by reflection. Incident light is represented with polarizations both parallel with and perpendicular to the plane of incidence.

incidence and reflection are close to Brewster's angle (56° for glass).

Polarization by reflection is not useful because the reflected intensity is already low (about 16% for glass at that angle). This can be improved upon by stacking many glass sheets. After six or so sheets most of the light is reflected and polarized (actually, as multiple beams). Polarization outside visible wavelengths can be done by this method, using appropriate materials that reflect at infrared or ultraviolet wavelengths. Quartz is a good ultraviolet reflector. Brewster's angle depends on the index of the material, of course.

10.4 BIREFRINGENCE AND POLARIZATION

Crystals that are optically anisotropic in their treatment of light make unusual changes in light passing through them. Crystals that polarize light are one example, and were known in the 17th century. A larger group of materials have different indices of refraction for different directions in the crystal. If the index is different in two directions, the material is *birefringent* (a term with the same root as and meaning the same as birefracting). The speed of light is different for waves vibrating in two different directions because the electrons behave asymmetrically, preferring to vibrate in one direction more than another. An unpolarized set of light waves entering such a crystal in certain directions is refracted in two directions, one for vertical polarization and one for horizontal polarization.

Birefringent crystals have one optic axis, which is a direction defined for the crystal structure and used to determine whether birefringence will occur. The optic axis is an axis of symmetry for the arrangement of atoms in the crystal. Birefringence is gone if light enters parallel with the optic axis, and it does not result in different directions for the rays if the light enters perpendicular to the optic axis. The rays still have different speeds and polarizations, however. Separation of the rays occurs when light enters at an angle to the optic axis (Fig. 10-5). For laboratory work, special crystals are marked to show the direction of the optic axis. Calcite is a common mineral with this property (and was once called Ice-

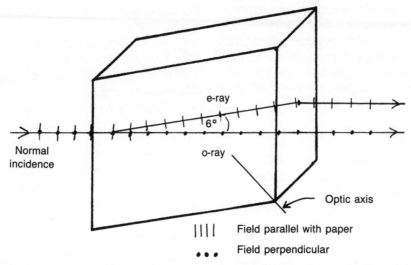

e-ray

6°

Normal
incidence

o-ray

Optic axis

|||| Field parallel with paper

••• Field perpendicular

Fig. 10-5. A birefringent crystal (calcite) separates randomly polarized light into ordinary ray and faster extraordinary ray, each polarized. The crystal is cut and used so that its optic axis is at an angle to the incident ray.

land spar). Other common birefringent crystals are tourmaline, quartz, ice, and rutile. In ice, the extraordinary ray moves slightly slower than the ordinary ray. Mica has three different indices, one for each direction, and therefore has two optic axes.

Experiment 10-3

If you have a crystal of calcite, place it on this printed page. Are there two images of the print? Rotate a polarizer in front of the crystal to see which image is affected. ∎

When light waves enter calcite normally, one set, now polarized, is called the ordinary ray and continues straight ahead as in Fig. 10-5. Its speed is determined by the ordinary index 1.658. The other set, with the other polarization, is the extraordinary ray and is refracted away from the normal at an angle of 6°. Its speed is determined by the extraordinary index 1.486. When it reaches the other face, it is refracted back to parallelism with the ordinary ray. By convention, polarization is often shown in diagrams by means of dots to represent field pointing out of the plane of the paper and short lines to show field polarized in the plane of the paper. The ordinary ray or o-ray obeys Snell's law whether or not the incidence is normal. The extraordinary ray

or e-ray does not follow Snell's law and does not even remain in the plane of incidence. It is not a transverse wave, although it does not differ greatly from these ordinary behaviors of light. The e-ray is different because the vibrating electrons contributing to it vibrate in somewhat different directions than the direction set by the incident light.

Several methods have been found to use birefringent crystals as polarizers of high accuracy. The goal is to separate the ordinary and extraordinary rays as far as possible over a broad band of wavelengths. The Nicol prism (after William Nicol in 1828) is constructed so that the o-ray is totally internally reflected into another direction, leaving the e-ray near the original direction (Fig. 10-6). The two parts of the prism are glued together with Canada balsam, which has an intermediate index of 1.55. The o-ray is absorbed by black paint on one face. The Nicol has a narrow field of view. Rays cannot differ much from normal incidence before failing to meet the condition for internal reflection.

The modern Wollaston and Rochon prisms use quartz or calcite to split light into o-rays and e-rays that are both usable. In the Rochon prism the incident light enters a prism parallel to its optic axis. The o-ray is never refracted so that it is achromatic (colors not separated). The Wollaston prism takes

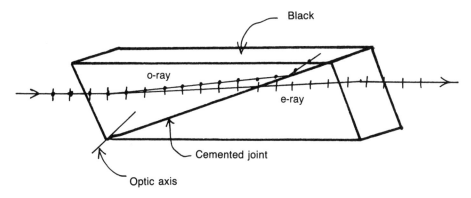

Fig. 10-6. The Nicol prism separates and preserves the extraordinary polarized ray.

light in normal to the optic axis and separates the rays more widely so that two usable but oppositely polarized waves are available. Many other polarizing birefringent prisms have been invented. All are fragile and expensive.

10.5 RETARDERS AND POLARIZERS

One polarization can be put out of phase with the other if advantage is taken of the different speeds of waves with different polarizations, such as occurs in a birefringent material. At a given wavelength, a certain thickness of material will introduce a certain phase difference. Half- and quarter-wave *retarders* or plates are made from calcite (but rarely), mica, and stretched sheets of polyvinyl (a polymer plastic). Cellophane wrap or tape, now becoming hard to find in consumer products, is a good, cheap retarding material. Plastic food wrap, once stretched, also works. If the retarder is a crystal, it is used with light incident perpendicular to the optic axis so that there is no separation of rays. A retarder such as the Babinet or Soleil compensator, made of two thin wedges of quartz, behaves as if it has variable thickness and therefore variable phase shift. At some position along the compensator, a particular wavelength will be retarded a set amount.

Experiment 10-4

Build up a stack of sheets of stretched plastic food-wrap until polarized light going through the stack seems unpolarized when viewed through a second, crossed polarizer. This will be a quarter-wave plate. ■

The phase added as light passes through a retarder can be certain multiples of a quarter or half cycle. The needed thickness d for a half-wave retarder can be calculated from the difference ($n_o - n_e$) between the indices and the wavelength thus:

$$d = (2m + 1)\lambda/2(n_o - n_e)$$

where m is any integer 0, 1, 2, etc. For a quarter-wave retarder, the 2 in the denominator is replaced with 4. Negative signs on d are ignored. A retarder is wavelength dependent and will properly retard just the wavelength it is designed for. If the light used is much different in wavelength, the polarization effects will be much less.

The principle purpose of a half-wave retarder is to change left circular polarized light to right, or vice versa (Fig. 10-7A). It also reverses the orientation of linear polarized light with respect to the "slow" axis, and it flips the orientation of elliptical polarized light. A quarter-wave retarder (Fig. 10-7B) changes linear to circular polarization and vice versa. This follows from the construction of circular polarization from two fields at right angles and 90° (a quarter cycle) out of phase. If the amplitudes are not equal, elliptical polarization is involved. In order that a quarter-wave retarder give circular and not elliptical polarization, incident linear polarization should be at a 45° angle to the slow axis of the retarder.

Circular polarizers are available in a one-piece combination with a linear polarizer whose axis is oriented at 45° to a quarter-wave retarder axis.

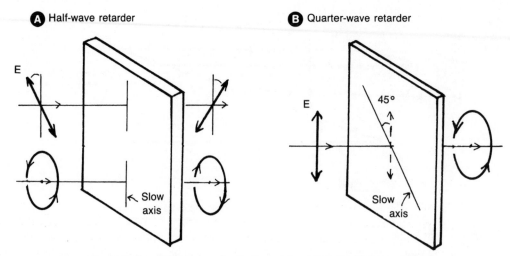

A Half-wave retarder   **B** Quarter-wave retarder

E

Slow axis

45°

E

Slow axis

Fig. 10-7. Two thicknesses of retarding plates using birefringent material.

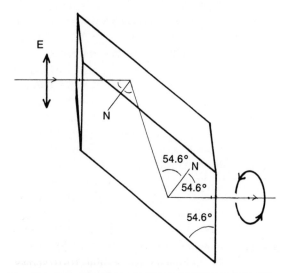

E

N

54.6° N

54.6°

54.6°

Fig. 10-8. A quarter-wave retarder made with internal reflection only (the Fresnel rhomb).

Light should enter the linear side. Whether one or two pieces, a circular polarizer can be used to determine if light has circular polarization and which handedness it has. Polarization can be analyzed with matrices. A series of polarizers and retarders can be represented mathematically and the final result predicted by multiplying matrices.

Reflection can be used for retardation. A more subtle result of the analysis of reflection in elec-

tromagnetic theory is the varying phase shift of internally reflected waves with orientations both perpendicular to and parallel with the plane of incidence. A variable phase difference occurs between waves with these polarizations. At an internally incident angle of 54.6°, glass causes a 45° phase difference. A 90° difference can be obtained by using two reflections at this angle. In the Fresnel rhomb made of glass, this effect is exploited to make a quarter-wave retarder as in Fig. 10-8. The input and exit faces are angled so that incidence is normal. After the first reflection, the light is elliptically polarized, a result of a 45° phase shift between the two orientations.

Thus far it has been assumed that polarizers are used on simple coherent light sources with definite polarizations. Monochromatic light necessarily has polarization. One electron vibrating (or many vibrating in concert) emits polarized light. White light comes from many electrons vibrating independently and has no such simple attributes. A polarizer can extract linearly polarized waves from white light, but they are not continuous wave trains, and they are of all colors and phases. For light with a small range of wavelengths (therefore approximately one color), the degree of polarization possible depends on the length of time a given polarization is expected. A given polarization state will last for a time inversely

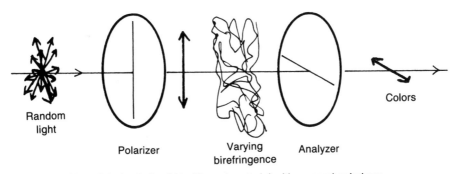

Fig. 10-9. Analysis of birefringent material with crossed polarizers.

proportional to the spread in wavelengths. Calculation of these properties is introduced in Chapter 13.

Because the amount of retardation depends on wavelength, retarders will produce assorted colors from white light. When substituted for a standard retarder, a test material whose birefringence and/or thickness varies can be analyzed with polarized light. Crumpled cellophane, mica, or almost any stressed plastic will show the effect. The colors are erroneously called "interference" colors, but they are actually a separation of colors by the variable light speed in the retarding material—a sort of dispersion without refraction. When the polarizers are crossed (at 90°), only the colors that suffer a rotation in polarization can pass through the system as shown in Fig. 10-9. This makes birefringent effects easy to see, as compared to aligned polarizers which produce a white background and transmit all colors except those rotated to extinction.

Experiment 10-5

Try various crumpled or otherwise stretched and stressed plastics between two crossed polarizers and observe the colored patterns seen. ∎

White light polarized and shone through a sample and a crossed polarizer results in colored bands seen in the sample, possibly resembling map contours. The practical applications are many. Because the background is dark, the stress patterns in transparent plastic are easily seen as colored contours. Models of ancient and modern buildings and mechanical parts are made of clear plastic and studied in this way. A measure of the amount of stress can be made by counting the number of dark bands or contours between a known unstressed region to a stressed region.

A phase-contrast microscope is designed to discriminate detail in transparent specimens by detecting the changes in phase of light as it passes through parts with slightly different indices of refraction. A retarder is part of the optical components needed to convert phase information into amplitude information.

10.6 POLARIZATION BY SCATTERING

Scattered light is polarized when the wavelength is comparable to the size of the particle or molecule which does the scattering. This occurs for infrared and visible light in many situations. What is important is the way the loosely bound electrons in the material vibrate when struck by the oscillating electric field of the incoming wave. As Fig. 10-10 shows, the electron can imitate the oscillation and vibrate to emit the same frequency and polarization that the incident light had. The electron provides a faithful copy only along the direction of incidence. At other angles, including overhead and below, the light is increasingly polarized. Light scattered 90° to the sides or above and below is plane polarized. If polarized light was incident, then there is no scattering above and below for vertical polarization and no scattering to the sides for horizontal polarization. The reason is simply that electrons cannot emit waves in exactly the direction they are vibrating.

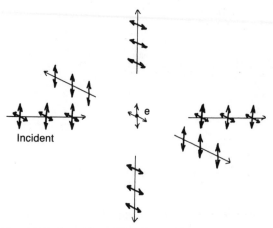

Fig. 10-10. Scattering of light by small particles polarizes it in various directions.

Experiment 10-6

Put a few drops of milk in a glass of water and mix well. Shine a flashlight through the glass in a darkened room. Through a polarizer used as detector, observe the light scattered from the milk solution from many angles around, above, and below. (In three dimensions, that is a lot of angles.) Repeat in the daytime on a clear day, looking at the sky (not milk), but do not look at the sun through a polarizer; only look to the sides. ■

In parts of the blue sky, the light is highly polarized. The light coming directly from sun to observer is not polarized. The light sent to the sides (east and west, up and down) is increasingly polarized. If the sun is in the south and you observe east through a polarizer, you are seeing blue light scattered sideways by air molecules to the east. A marked polarizer will reveal the light to be partially vertically polarized. When rotated to the horizontal, the polaroid will appear darker. Polarized skylight is the explanation for the remarkable navigating ability of insects, birds, and other animals. Their eyes can detect polarization accurately and therefore tell time of day and/or direction. The Vikings made rare use of the polarized sky to navigate, detecting polarization with a crystal not definitely identified. You may be able to see sky polarization without a polarizer. It appears as a faint yellow pattern called Haidinger's brushes. Rainbows are polarized, too, because the reflection inside water droplets occurs very near Brewster's angle, eliminating one polarization from random sunlight and preserving the other.

10.7 OPTICAL ACTIVITY

Optical activity refers not to just any work with optics but to a special effect that certain materials have on linearly polarized light. In quartz, for example, the plane of polarization rotates in space as the wave moves through (Fig. 10-11A). This was discovered in 1811 by Dominique Arago. It even occurs in liquids. The direction of rotation can be right or left, and the right-hand rule tells which name to give. Rotation following the curled fingers of the right hand when the thumb points in the direction of propagation is called dextro-rotation or d-rotation. The other way is levo-rotation or l-rotation. The rotation occurs over substantial distance (millimeters) not on the scale of wavelength. Quartz rotates the plane of vibration at the rate of 21° per millimeter.

Experiment 10-7

Put some sugar syrup in a glass and shine polarized light through it. Examine the emergent light with an analyzer. Rotate it to see what colors emerge at different polarizations. ■

Because plane polarized light can be composed of both circular forms, circular birefringence is an explanation for optical rotation. Some materials have different light speed for right-verus left-circular polarization; the twist of the molecules encourages one hand of propagation over the other. One gets ahead of the other, and the combination results in a rotation of the equivalent plane polarization. The silicon and oxygen atoms in quartz are known to be connected in left- or right-helical patterns. The same effect has been demonstrated with microwaves passing through a box of springs all coiled the same way!

Optical activity indicates anisotropy in materials, but it may occur when there is no birefringence. Liquids provide smaller rotation that depends on the concentration. The rotation depends also on wavelength, and so there is rotary dispersion. Blue light waves are rotated more than red over the same distance. The rotation is due to the inherent left- or

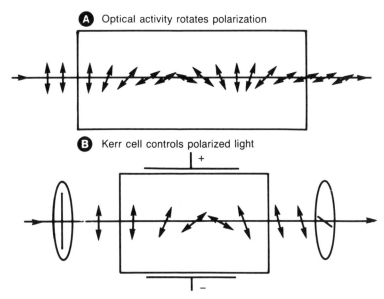

Fig. 10-11. An optically active material causes linear polarized light to rotate over distance. When used in a Kerr cell, light can be switched on and off electrically.

right-handed complex molecules in the substance or solution. Optical activity is a test of both the handedness of the substance and the concentration of it. This is especially useful with medical and biological specimens (for example, urine). The majority of molecules synthesized by and used by life have helical form (l-rotary for most and d-rotary for sugars). Organic molecules synthesized randomly by technological processes have equal amounts of l- and r-rotary forms. Life elsewhere could just as well be based on r-rotary molecules. Equal mixtures of l- and r-rotary forms are not optically active because the effects cancel.

10.8 OPTICAL MODULATION

The polarization state and other properties of light can be affected by external influences as well as by the materials through which the light passes. Because light consists of electric and magnetic oscillating fields, one would expect to find technological means of affecting the fields and the rate of oscillation. In the Faraday or *magneto-optic* effect (from Michael Faraday in 1845) a strong magnetic field can rotate the polarization, not by affecting the magnetic field in light but by affecting the environ-

ment through which the electric field travels. The field should point in the direction of travel and is obtained by winding a coil around special material such as YIG (yttrium-iron-garnate). The external field changes the behavior of the electrons in the material. This time, however, the speeds of right and left circularly polarized light are affected, not the speeds of linear polarization. The size of the effect depends on the field strength, the distance traveled, and the material.

The effect is very weak in air. In water in the weak field of the Earth about 2° of rotation occur in 1 centimeter. The handedness of the rotation depends on the material and the direction of the field. The Faraday effect can modulate or vary the polarization of a light wave if a material with a strong effect is used in a strong field. The modulation rate can be very high and is therefore useful for transmitting high-speed messages. YIG is used in the infrared where it is transparent. In the different Cotton-Mouton effect the magnetic field is perpendicular to the light travel and causes birefringence, which then rotates the polarization of light.

The *electro-optic* or Kerr effect (from John Kerr in 1875) uses an electric field to make an isotropic material birefringent, thereby affecting the polarization. The difference between o-ray and e-ray speeds depends strongly on external field strength as well as on wavelength and the material. A Kerr cell (Fig. 10-11B) exploits the effect to make a modulator or shutter. The material is a liquid hydrocarbon (hazardous ones like benzene, chloroform, nitrotoluene), or possibly water or a crystal. Without field applied, the crossed polarizers permit no light to pass through. With the field, some amount of rotation occurs as if the tank of liquid were a controlled retarder. The light beam can be switched on and off in less than 10^{-10} second, very useful for studying fast phenomena. Slower Kerr cells were first used to measure the speed of light in 1925, still operating much faster than mechanical beam choppers.

The Pockels effect (Friedrich Pockels, 1893) occurs with certain kinds of crystals. Again, an external electric field influences birefringence. When the field is on, light is retarded in the Pockels cell and can pass between two crossed polarizers. The response time as a shutter is slower, perhaps 10^{-8} second. It can produce controlled polarization or controlled intensity. The search for better materials for Kerr and Pockels cells continues into the ever more exotic (for example, lithium tantalate). Solid materials are desirable, and a larger effect can be obtained by putting a reflector at one end of the crystal so that the light passes twice through the same cell.

10.9 LIQUID CRYSTALS

Liquid crystals are an unusual form of matter that flow like a liquid but are ordered like crystals. Their reflectance or polarization is easily controlled by small electric fields so that they are useful as passive display devices. No power need be expended on emitting light. Display occurs by means of reflected light when the electric field makes the liquid crystal diffusely reflecting (and darker). Figure 10-12 shows a liquid crystal display used in transmission. No character is seen until optical activity is switched on for selected segments so that light is rotated and can pass the crossed polarizers.

The physical chemistry of liquid crystals is complex. There are three basic types (nematic, smectic, and cholesteric), and they are rod-shaped organic molecules sometimes packed in layers. When the layers have a helical arrangement in the cholesteric type, rotation of polarized light results, and there is circular birefringence. The optical activity is much stronger than in quartz. The birefringence results in an iridescent appearance in white light. Nematic and smectic types have linear birefringence, or a polarizing dye can be mixed with the nematic molecules when a polarizer is wanted. The cholesteric type has color changes very sensitive to temperature and thus has been used in passive digital thermometers. Smectic types have been found subject to the electro-optic effect and can switch light on and off very rapidly.

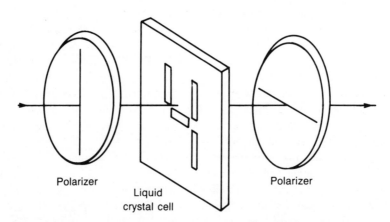

Polarizer

Liquid
crystal cell

Polarizer

Fig. 10-12. In the transmission mode, a liquid crystal character is visible when an electric field causes molecules in selected segments to rotate initially polarized light to pass a second crossed polarizer.

11

Quantized Light

Light modeled as particles was once used to attempt to explain the many diverse optical phenomena observed before the 20th century. Then wave theory culminating in Maxwell's electromagnetic waves took over. But a return of a particle theory for light was necessary before the scientific and technological revolutions of the 20th century could develop. This chapter is also a logical place to discuss the radiation and energy properties of light, a field called radiometry. Some of many units of measurement are explained.

11.1 PHOTONS VERSUS WAVES

In 1905 the particle model of light returned when Albert Einstein (in one of three revolutionary papers in three different fields that year) proposed the photon of light as an explanation of observations that had puzzled physicists for many years. He studied the photoelectric effect, where ultraviolet light was found to eject electrons from metal surfaces. Whether electrons were ejected or not did not depend on how intense the light was. Only the wave-length or frequency mattered. If light were waves, as in Fig. 11-1A, one could simply bathe the metal with more intense waves until an electron had enough energy to break free of the surface. Instead, it was discovered that very weak light of certain frequencies would eject electrons (Fig. 11-1B). There must be a relation between the frequency of light and the energy it carries. Moreover, the light energy is delivered all in one chunk or *quantum* to the electron. The supposed ultraviolet light wave, heretofore spread over the metal surface, suddenly acted as if it were a single particle called a *photon* (Greek for light) and delivered right to the doorstep of a single electron.

Physicists had been noticing, as Hertz did in 1880, the effect of light on electrons leaving metals. They had even been able to control carefully the energy needed for an electron to leave and relate it to the frequency of the light used. Provided that the light was sufficiently energetic (not the same as wave energy) to make a few electrons leave metals, then the number ejected was directly proportional to the intensity of the light used. Wave effects also

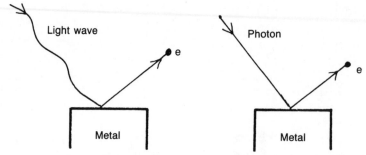

Fig. 11-1. The photoelectric effect.

occurred. Einstein's contribution was to show that the energy E picked up by an electron was in the amount:

$$E = hf$$

where h is Planck's constant $6.63(10)^{-34}$ joule second (in SI units), and f is the frequency of the light used. (For the present E denotes energy rather than electric field).

Max Planck had proposed earlier that light carried energy proportional to frequency. This was an attempt to explain certain other experiments where light seemed to exist in discrete chunks or quanta. Maxwell had already proven with electromagnetism that the energy and momentum p of a wave was related by $E = pc$. Momentum can be defined here as a quantity carried by any moving object that represents its direction, its speed, and the impact it can have. The amount of momentum is the product of speed and mass ($p = mv$).

Photons travel at the speed of light and are massless. Einstein's relativity requires that such particles be massless, or else their speed would have them carrying infinite energy. Yet photons deliver impact when they strike anything. Photons must have a definite and specially defined momentum. The momentum is related to the wavelength by a corresponding relation:

$$p = h/\lambda$$

The impact that a wave or photon would impart to an object it struck is inversely proportional to its wavelength, again through Planck's constant. That this is really mechanical momentum is demonstrated by the possibility of light sailing, as well as by delicate laboratory experiments that measure the pressure on a surface due to strong light. The radiation pressure P (a force per unit area) can be calculated from the light intensity (a power per unit area) by dividing by the speed of light:

$$P = I/c$$

Alternatively, the pressure of photons striking a surface is found by dividing their momentum by the area and the time interval between successive photon hits ($P = p/At$). The results are equivalent, and both apply to the case of complete absorption of waves or photons. If the surface is perfectly reflecting, then the pressure is twice as great. The surface receives double impact if a wave or photon departs in the same direction from whence it came.

With a power of ten of the order of 10^{-34}, h is clearly a very small number. A typical red light frequency of $4(10)^{14}$ is equivalent to about 10^{-19} joules of energy. This sounds like very little, but the single electron carrying this energy is moving at a high speed and represents a high temperature (over $1000°C$). A metal must be heated to glowing to give each surface electron this amount of energy. But, very sensitive instruments are needed to detect single visible photons. The eye needs to receive several in a small region before a flash of light is detected.

Physicists faced quite a dilemma in the early 20th century. Electrons clearly were knocked out of metals by photons, yet an enormous body of knowledge in wave optics had accumulated. Specifically, light waves could divide into parts in order to pass through two slits, then form a very wave-like interference pattern (Fig. 11-2A). A photon should choose one slit to go through and land in a small pile associated with that slit (Fig. 11-2B), not spread out into a series of fringes all up and down the screen. This is the dual behavior of light. Experiments proving the quantized nature of electromagnetic radiation were often done outside visible wavelengths. Only recently has technology advanced to support more quantum experiments with visible light.

A resolution of any apparent paradoxes about light came when it was discovered that a single photon would act as if it interfered with itself in passing through the two slits. Light can be made so dim that one photon at a time goes through the experimental setup. Where these photons land seems almost random, at first, and unrelated to the two slits. As photon count increases, the piles of photons on the screen form a pattern exactly the same as wave theory predicts. A single photon is unpredictable, usually landing near a maximum but sometimes in a minimum. A multitude of photons form a precise pattern even if they interfere one by one. More astonishing, but in keeping with a sort of symmetry to physical theory, is the experimental proof that nature also works the other way around. Objects

hitherto thought to be purely particles, including good, solid, heavy ones like protons and neutrons, were found to behave like waves when given a chance to diffract and interfere. All matter is dual. So light will never be determined to be either just waves or just photons. It is ''something else'' that manifests as waves or photons depending on the situation.

The ''interference'' of single photons has been put to good use recently by astronomers seeking better resolution of very dim objects. In *speckle interferometry* the positions of individual photons are recorded electronically after they arrive through one or more large telescopes from an object. The photons come too infrequently for film exposure, but their origin can be pinpointed after data have accumulated on many photons.

11.2 BLACKBODY RADIATION AND LIGHT INTENSITY

Before the consequences of light as photons are fully considered, more background is needed on the nature of light as radiation on the large scale rather than microscale. No waves or photons need be considered. As soon as good dispersive instruments were available, the characteristics of the light emitted by a glowing body were studied. The body could

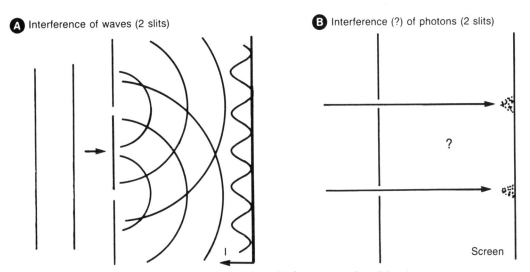

Fig. 11-2. Interference from two slits for waves and particles.
Photons do not pile up behind each slit but instead interfere like waves.

be the sun, a glowing filament, a candle, a mercury lamp, or any of many other bodies. Improving detectors allowed the intensity of light to be measured in ever shrinking bandwidths—that is, smaller ranges of wavelengths. The width of the spectrum studied also grew. William Herschel, for example, discovered the existence of infrared in the spectrum in 1800 by putting a thermometer beyond the red end of the spectrum from a prism. For glowing objects like the sun or a candle, a characteristic graph or curve was found (Fig. 11-3) when intensity per wavelength (spectral flux) was plotted as a function of wavelength.

A glowing object has many atoms independently emitting light of many colors and absorbing some, too. Its radiation behaves as if it were a perfectly absorbing blackbody as well, hence the name of the curves. A law named after Gustav Kirchhoff states that a material emits radiation as well as it absorbs it at any given wavelength. It is difficult to envision the fiercely glowing sun as a blackbody, but the name is significant to optical scientists and technicians. The curves shown are for bodies of widely differing temperature. Radiation temperatures are usually measured in degrees absolute (denoted by K with or without °). The sun has a peak in the middle of the visible range, emitting the most radiation in the yellow. The sun is called a yellow star. It looks white to us because the amount of blue and red nearly balance as the curve shows.

The candle has a peak in the infrared. More heat energy is felt than light energy seen from it. The peak wavelength is inversely proportional to the temperature (Wien's law), so the candle with approximately one-sixth the temperature of the sun has a peak wavelength six times longer (one-sixth as energetic). Even if the candle were as big as the sun, the intensity from the candle at any wavelength would be much less than that from the sun because cooler objects emit less radiation at all wavelengths. The relevant radiation law (Stefen-Boltzmann law) states that the spectral flux is proportional to the fourth power of the temperature. Thus the curves are nested and never cross.

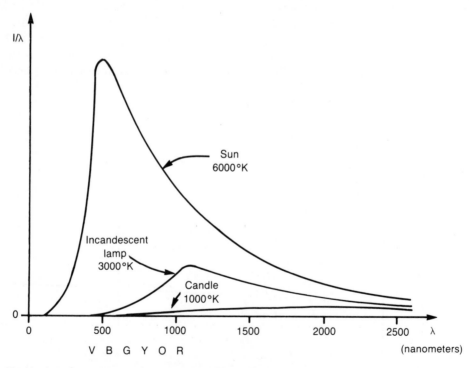

Fig. 11-3. Blackbody radiation curves for hot objects at several temperatures. The vertical scale showing spectral flux is compressed (quasilogarithmic).

In contrast to blackbody radiation curves for bodies with randomly moving and emitting atoms in thermal equilibrium, some sources of light are nonequilibrium and favor some wavelengths over others. A fluorescent light, for example, emits light strongly at certain wavelengths characteristic of the hot mercury vapor inside, together with a relatively smooth spectrum from the phosphor coating on the tube (Fig. 11-4). How all this works will be clearer when atoms are discussed. A good laser emits a very narrow band, almost a line of radiation. It is as far from equilibrium as one can get.

Late 19th century physicists such as Lord Rayleigh, Sir Jeans, and Max Planck were interested in predicting the shape of the radiation curves from electromagnetic theory and concerned about what happens down in the left corner of the curve. The region near zero wavelength represents very high frequency and very energetic light (ultraviolet, x-rays, gamma rays, and beyond). A glowing candle should not be emitting much of this sort of thing. An explanation seemed impossible until the advent of quantized light.

11.3 PHOTON AND WAVE RADIATION MEASUREMENTS

The variables and units used to describe light intensity and related energy aspects are many, varied, and often archaic despite wide use. The term intensity used liberally in this and other books is not as easily defined as one might think. The basic physical quantity is the *radiant intensity*, defined as the power coming from a point source passing through a unit of solid angle. Rather than try to visualize and calculate a solid angle, the equivalent form for radiation passing through a spherical surface around the source is given here. A spherical surface of radius 1 has 4π or about 12 units (steradians) of solid angle. Intensity varies with distance from the source, which is why physicists prefer a definition that avoids distance effects. When attention is directed to detecting light energy, the detected variable is *irradiance*, an intensity defined as power received per unit area. The ending "-ance" is the cue that the energy is measured as received (detected). Whether waves or photons, the light received from a source decreases with distance according to an inverse square law. For example, if the distance from the source doubles, the intensity or irradiance decreases by a factor of four.

Radiation and luminosity have a basic difference. *Radiation* is that which can be measured with instruments that detect energy and refers to any part of the electromagnetic spectrum. *Luminosity* refers

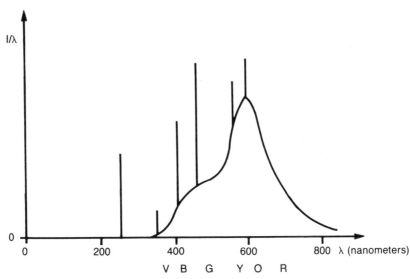

Fig. 11-4. A nonequilibrium spectral curve for an ordinary fluorescent lamp, showing broadband emission from the phosphor and narrow lines from the mercury.

to visible light and is affected by how well the eye perceives various colors. Radiant intensity might describe the heat received by a detector from a hot source. For example, the maximum irradiance in sunlight at the Earth's surface is about 1,000 watts (W) per square meter. Luminous intensity measures what the eye sees—so many watts per square meter of bright yellowish light—and ignores the substantial infrared. The connection between radiant and luminous units is standardized at 555 nanometers (yellow-green). At other wavelengths, the measured luminous intensity is adjusted to be less to match the weaker response of the average eye.

Luminous intensity is measured in *candelas*, which have little to do with candles. A candela is defined as that radiation emitted visibly by a blackbody of a certain size at a certain temperature. *Luminous power* is the portion of emitted power that can be seen by the eye. It is measured with the *lumen*, a unit derived from a luminous intensity of one candela times one unit of solid angle. An older unit is the standard candle which is about 0.02 W of visible light. Incandescent lights emit only about 3% of the input power as visible light. Thus 1 meter from a 100 W light bulb, about 1/4 W per square meter is detected as visible light. Lighting efficiency for human use is compared by using the ratio of lumens of luminous power emitted to the watts of electric power consumed. A typical incandescent 100-watt bulb gives 17 lumens per watt, whereas the maximum theoretically possible is 220 lumens per watt.

Brightness is a subjective response of the eye and brain to luminosity. It is a luminous density because we commonly examine a small area of a light source in comparing brightness. The visible moon sends an enormous amount of radiant energy toward the Earth, but we see it as no brighter than a dim light bulb. As shown earlier, the brightness of an image can be increased or decreased as the image is contracted or expanded in size.

The official SI unit for light is the *lux* (Latin for light), which measures the *illuminance* on a surface. That is the power received on a unit of area and equivalent to a lumen per square meter. A footcandle is an outdated unit of illuminance. Laboratory light meters may read in lux or the outdated footcandle. Readings in lux assume that 555 nanometers is the only wavelength used. The best meters read in watts per square meter. This represents the energy received from the sun or any other source and is readily interpreted as heating power. Camera light meters, if they read in a photometric unit at all, will use footcandles which can be converted to modern units (1 footcandle = 10.76 lumens per square meter). More detail on radiation and luminosity units is found in references such as Meyer-Arendt.

Some common light levels (illuminances) are given here to familiarize the reader better with light units. Direct sunlight at noon in the summer is about 100,000 lux. In the shade, the illuminance is about $\frac{1}{10}$ that value. A camera exposure meter registers this as a large change, but the eye adjusts so that we hardly notice it. At twilight after sunset, the level has dropped to about 10 lux. The full moon provides about 0.1 lux, and the moonless rural sky is at about 0.001 lux. (Note the variation of about 10^8 in natural light levels.) Light levels in buildings depend on what standards are used.

Until the first energy crisis, light levels had been rising—e.g., over 1000 lux for classrooms and offices. Then it was discovered these levels could be halved with no difficulty. Usable home lighting varies from about 500 lux down to 50. The light needed for close-up work is as bright as daytime shade, about 10,000 lux. The eye cannot adjust to reading in direct sunlight, but this light is not too strong for short-term fine work such as examining the details of printed letters. Commercial lighting is a quandry because a high level is needed for reading product labels and so forth, but otherwise the level is rather high for casual vision and requires much electric energy. Classroom desks are lit from far above for reading fine print, whereas at home we can use less than $\frac{1}{6}$ the luminous intensity from a small lamp set nearby. In either case, the illuminance on page might be about the same.

The intensity I of a light wave can be converted to photons if only one frequency is present. The number of photons passing through a unit of area in a unit of time (the photon flux density) is given by I/hf. Any detector of photons has an area, which

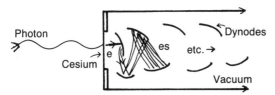

Fig. 11-5. A simplified diagram of a photomultiplier that can detect one photon and convert it into a cascade of electric current.

is to be multiplied by the photon flux density to obtain the total photon count per second. The detection of small numbers of photons is done with a photocell. It makes use of the photoelectric effect, in which photons more energetic than a certain minimum knock electrons out of a suitable metal. Red light can be detected with metals such as cesium, but with poor sensitivity. To detect as few as one photon, a photomultiplier is used whereby the first loose electron is used to knock more out and so on through several stages (Fig. 11-5). Somewhat larger amounts of radiant energy can be detected with a calorimeter, which at its most sensitive measures the rise in temperature of a tiny piece of material in a liquid helium cold chamber. Weak infrared is detected by a germanium bolometer. Recently a solid state silicon avalanche photomultiplier has been developed that is more sensitive than the vacuum tube type if kept cold. It works over the visible and infrared and detects single photons, whereas the vacuum type cannot detect infrared. Because it is a microcircuit device, arrays of ultra-sensitive detectors can now be built for astronomical or laboratory use.

11.4 UNCERTAINTY AND PHOTONS

A major discovery in 20th century physics is that all matter and energy has somewhat uncertain results in basic measurements such as speed, time, energy, and position. The situation is serious at the microscale of photons and electrons but virtually undetectable at the human scale. Planck's constant h is at the heart of the uncertainty and sets the sizes where strange things happen. The *uncertainty principle* states that the accuracy with which one can know the position and speed of a tiny particle is a trade-off. The better the position is known, the less

can be learned about the speed, and vice versa. The sizes of the errors allowable with each, when multiplied together, give a value approximately equal to or greater than Planck's constant. The same holds for knowledge of the energy versus the amount of time needed to measure it. More correctly, speed must be represented by momentum. In symbols the uncertainty principle has the form:

$$\Delta p \Delta x \stackrel{\sim}{>} h$$
$$\Delta E \Delta t \stackrel{\sim}{>} h$$

where the symbol Δ (Greek capital delta, spoken simply delta) denotes that there is error or uncertainty in the quantity following it.

A relation between the uncertainty of which slit a photon goes through and the sizes of the slits needed to show interference should be apparent. To discern which slit a photon goes through, its position needs to be pinned down to no more than the distance between slits. Then the uncertainty principle and Planck's constant require that the uncertainty in momentum be larger than some tiny number. When calculations are completed, this turns out to be equivalent to the size of the wavelength used. This uncertainty agrees with the results of wave interference. When light of that wavelength is used, the resulting pattern is spread all over the screen, and one cannot determine which slit the wave or photon went through.

Another practical application of the uncertainty principle is the light microscope. Suppose that one wants to make out details as small as 10^{-7} meter with a perfect microscope. Then the light used must have a minimum uncertainty in momentum of 10^{-26} momentum units (Planck's h was rounded off to 10^{-33} for this estimation). From the special energy-momentum relation for light ($E = pc$), the uncertainty in energy can be estimated from the uncertainty in momentum. This gives about 10^{-18} joules (in round powers of ten). Converting to frequency (with $E = hf$), the result is a minimum uncertainty in frequency of about 10^{15} Hz. This is near ultraviolet, just as expected from the postulated sizes of interest.

The key to seeing smaller detail is smaller wavelength or higher energy photons. The electron microscope uses the next step down—fast electrons that are equivalent to mild x-rays in resolving power but much easier to focus an image with. Using particles besides photons to study small objects is still called optics, but is beyond the scope of this book. Ways are now being developed to use light to resolve objects only a fraction of a wavelength in size, but there is nevertheless a lower limit—perhaps a hundredth of a wavelength. Smaller than that, resolution will be impossible.

Assuming that the idea of photons diffracting and interfering is comfortable, what about reflection, refraction, lenses, and dispersion? Reflection of photons seems obvious. They should bounce. But no surface is smooth at the scale of a particle with infinitesimal size going at light speed. A smooth metal mirror would appear as a seething, bumpy, vibrating maze of electric fields. The photon needs to act like a wave to reflect specularly from this mess. Similarly in refracting materials like lenses, photons have a collision-ridden trip. There is no simple way to show that every photon is slowed to a certain speed, and the explanation is best left to the wave model. Photons cannot look ahead and know which way they should change course after leaving a surface. Light waves in some way sense out the space ahead.

Light, whether waves or photons, is inherently noisy. There are unavoidable random quantum fluctuations of amplitude and phase for light waves, no matter how pure, and fluctuations of photon count. A way has been found to "squeeze" some noise out of light, so that it is quieter than the dark, which also has quantum noise. The uncertainty principle is not violated, but instead one aspect of it is employed at an advantage over another. The experimental methods can transfer noise or uncertainty from the amplitude to the phase, for example. Reduction in noise by a factor of 10 or more is expected and will permit a similar improvement in in-terferomic measurements of position.

11.5 POLARIZATION

What about a connection between polarized light and photons? How is one to explain polarizers and circularly polarized light? The answer is easy if one accepts that each photon carries a bit of rotation called *spin*. This spin can be oriented along the direction of travel, giving left circularly polarized light, or oriented backwards for right circularly polarized light. Photons cannot have any other orientation. Unfortunately, this link between the handedness of light waves and the spin of photons is inconsistent with a right hand rule, but no change is likely. Light that is unpolarized consists of about equal numbers of right and left circularly polarized waves or of left and right photons. Linearly polarized light consists of exactly equal numbers of each. A single photon cannot be forced to show linear polarization but instead will be measured to have spin either forward or reversed (left or right). The size of the spin carried by photons is quantized in the amount of $h/2\pi = 1.06(10)^{-34}$ in SI units.

In the form of photons, polarized light used in certain experiments raises some very disturbing questions about the laws of nature. In one experiment, certain particles emit two photons in opposite directions, each with opposite spin. These are allowed to travel a long way and then each are detected in a polarization detector. As soon as one is detected, its spin can be measured, and the spin of the other photon is then known without even using the detector. But this result seems to violate one or another basic principle. Either information is sent faster than the speed of light from one detector to the other, or photons are not simply points of energy and contain an internal mechanism that "knows" what spin the other has. See references on quantized matter for more on this Einstein-Podolsky-Rosen thought experiment.

12

Atomic Theory and Light

Just about all the exciting behaviors of light involve atoms, the building blocks of all materials. Light starts from atoms, is changed by them, and is soaked up (absorbed) by atoms. You cannot see light without atoms. Atoms deserve more study and appreciation than given thus far. This is not a book on atoms or quantum theory, so only the amount of introduction to atoms necessary to support discussions of light is given. Almost no mathematical description can be given at this level.

12.1 MODELS OF THE ATOM

Two models of the atom are offered here, both much simplified compared to what atoms seem really to be like. One is the Bohr model (from Niels Bohr about 1913), which resembles a miniature solar system model with a center or nucleus like a sun and a set of orbiting electrons like planets (Fig. 12-1). The analogy is strained at best, and an important modification is based on the discovery that the electrons cannot pursue just any orbits but are limited to certain ones. Each orbit represents an energy.

The farther out from the nucleus, the higher the energy of the electron. Only certain values of energy are permitted in both the Bohr model and modern models. The tendency of orbiting electrons to emit radiation continuously was one reason the arbitrary restriction on orbits was adopted. Otherwise all atoms would collapse as the electrons radiate away energy and spiral into the center.

Almost all the weight (really, mass) of the atom is at a tiny place in the center called the *nucleus*. Electrons may be infinitesimal, but they are light and move around a lot, managing to occupy most of the vast space around the atom. The whole package has a size between 10^{-10} and 10^{-9} meter. Different kinds of atoms are built up by adding positive charges (protons) to the nucleus and an equal number of negative charges as electrons. The electric force between positive nucleus and negative electrons is very strong. The electrons close to the center are tightly bound and do not participate in optical effects. It takes x-rays to reach them, and their energy is large. The energy of bound electrons is defined to

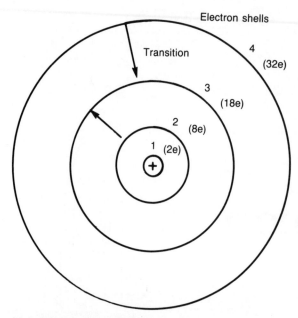

Electron shells

Transition

4 (32e)

3 (18e)

2 (8e)

1 (2e)

Fig. 12-1. The Bohr model of the atom, showing the maximum electron content of four orbits or shells to scale. Complex substructure holds each electron in a distinct state, and the number of positive charges in the nucleus is the same as the number of electrons. For hydrogen, the one electron resides normally in the lowest (number 1) shell.

be negative so that outer lightly bound electrons are at a higher energy level than inner electrons. Atoms tend to be neutral and have equal amounts of each kind of charge. If an atom were not neutral, an electron would soon arrive or depart to make it so. Atoms bind to each other via the electric force to form molecules, crystals, metals, and so forth. There are about 100 different kinds of atoms, each representing a distinct chemical element. Some materials of optical interest are pure elements consisting entirely of one kind of atom, but most are compounds of different elements.

The energy arrangements in an atom become very complex when there are more than a few electrons. Keeping track of the electrons in (fictitious) orbits becomes hopeless. The other atomic model is more abstract. The atom as a whole is thought of purely in terms of energy. It can possess various quantities of energy. Energy is measured for atoms in terms of electron volts more often than in joules. An *electron volt* (eV) is the energy acquired by an

electron when pulled on by the electric field present between two plates with a difference of one volt. To us this unit is small, $1.6(10)^{-19}$ joules, but an electron with 1 eV is already pretty hot, about 10,000°K (hotter than the surface of the sun).

The energy *level* or *state* diagram shown in Fig. 12-2A is for hydrogen, the simplest atom. It has one electron which, when closest to the nucleus of one proton, has an energy of −13.6 eV. This is state number 1. If energy is put into the atom by shining the appropriate light on it, the electron can gain 10.2 eV and move up to state 2, where it has −3.4 eV (Fig. 12-2B). Any atom with at least one electron moved from a lower state is called *excited*. This process can continue, with the electron (really, the whole atom) absorbing smaller and smaller amounts of additional energy as it climbs to higher states. There is a limit when a total of 13.6 eV has been put into hydrogen. The electron is now free of the atom and can move away. The atom is *ionized* when it loses an electron. Perhaps it is clear now why the bottom state 1, called the *ground state*, is defined to be −13.6 eV.

In a reverse of the ionization process, consider an electron resting near a hydrogen atom without one (Fig. 12-2C). It can drop into one of the states available. If the state is number 3 at −1.5 eV, the excess energy of the electron is emitted as light of energy 1.5 eV (actually a little more). This can be converted to a color or frequency of light (by $E = hf$) and is found to be infrared. After a short time, the electron realizes that having lower energy is more desirable and drops to another state, say number 2 at −3.4 eV. It must radiate the difference −1.5 − (−3.4) = 1.9 eV (red light). Finally, when it drops to the ground state, −3.4 − (−13.6) = 10.2 eV must be released, and this comes out as fierce ultraviolet.

Upon applying units conversions to Planck's relation of energy and frequency, the useful result is obtained that the wavelength (in nanometers) emitted is equal to 1240 divided by the energy emitted (in eV). Planck's law of radiation is stated mathematically as:

$$hf = \Delta E = E_f - E_i$$

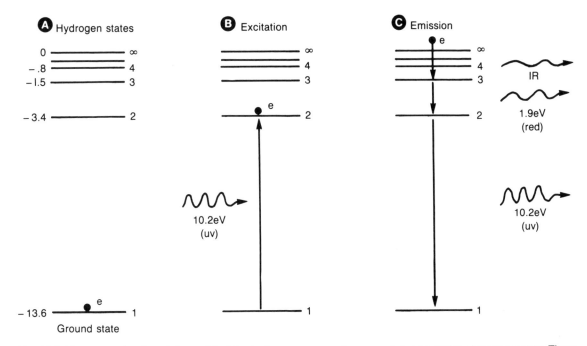

Fig. 12-2. The energy level or state model of the hydrogen atom and some radiative transitions between states. There are many states narrowly spaced just below the ionization level marked ∞.

where subscripts *f* and *i* denote the final and initial energy states.

All atoms have many energy states, and these are a very complex set in atoms with many electrons. A goal of *spectroscopy* is to learn the energy state structure of atoms by analyzing the various wavelengths of light the atoms can be made to emit. A principle tool is the diffraction grating spectroscope. Detailed measurement is done with interferometers. The energy difference between any two states is very precise, and an almost monochromatic light wave is emitted. Such a narrow band of color or wavelength is called a spectral line because it looks like a colored line with a spectroscope when the source of light is a narrow slit. When line wavelengths are measured to eight decimal places, the number of distinguishable lines is very large and those peculiar to each kind of atom can be identified.

Simple hydrogen has many spectra (Fig. 12-3). These arise from the many ways that electrons can jump down one or more states. The spectra come in families corresponding to the possible jumps down

to state 1 (Lyman series), to state 2 (Balmer series), to state 3 (Paschen series), and so on. Hydrogen atomic behavior is almost completely understood, since a quantized theory of the hydrogen atom predicts the spectrum, including details, to high accuracy. Helium and more complex atoms have much more complex spectra. Many of the energy levels consist of pairs or triplets of closely spaced lines. When an electron in sodium changes state from either of a pair of states to ground (Fig. 12-4), one of two closely spaced yellow lines is emitted. These two lines (a doublet) differ by only 0.6 nanometer. In consolation, many of the possibe transitions indicated by a state diagram are not permitted by other laws of physics.

Order is brought to the many energy states and the corresponding spectral lines by realizing that several behaviors of an atom besides its energy are quantized. The electrons can have various amounts of rotational momentum (*angular momentum*) around the atom, and this quantity can occur with only certain values. Such rotational states exist even

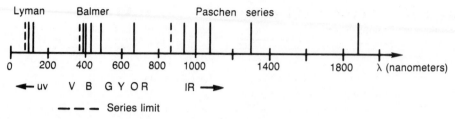

Fig. 12-3. Three series of spectra for hydrogen. Because of the scaling of wavelength, many lines seem to bunch together in the ultraviolet.

if the atom is not considered with the Bohr model but rather with more abstract and accurate models. The magnetic field within the atom itself alters the energy of the rotational states so that they are different, depending on the amount of rotation. The spin of electrons affects the energy states they can be in. Altogether, every state of the atom is distinct, and every electron must have a different energy, if only slightly. When an electron returns to an atom, it will drop only to an unoccupied state. Visible light is in the energy range where atoms are at most only partially ionized, missing a few electrons.

Experiment 12-1

If you can borrow, or buy from a consumer science supply company, a replica diffraction grating, look at various light sources to find lines. The source of light must be narrow, so the grating should be assembled in a simple spectrometer. Mount the grating at one end of a cardboard tube that has a narrow slit at the other end. Look at fluorescent and incandescent lamps, highway lights, and salt sprinkled in a flame, but not at the sun. ■

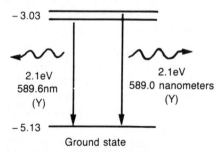

Fig. 12-4. Simplified energy state diagram for sodium, showing the transitions for the pair of prominent yellow lines.

12.2 DETAILS OF THE EMISSION OF LIGHT BY ATOMS

Every kind of atom (chemical element) has a characteristic spectrum. Helium was first discovered in the sun by accounting for the spectral lines of other known elements in the sun. All or most lines are seen in a spectrum of a given element because very many atoms participate in emission. When the source is hot, or excited by electrical discharge, atoms are randomly affected so that electrons are promoted to a variety of states or torn away entirely. Nearly every permitted transition occurs, so that most permissible spectral lines are seen. The bright double line from sodium is seen because similar numbers of atoms are changing state from each of the two close states to the ground state at any time.

On the macroscopic scale, atoms are made to emit light by making them hot. Heat is conveyed to atoms by collisions with other atoms. In a gas, this is the only way to pass the heat around. Atoms are caused to move at high speed by direct contact with hot atoms in a flame. Or electric force accelerates ionized atoms to hit more atoms and knock electrons from them, ionizing them. A collection of ionized atoms is needed if the fullest spectrum is to be observed. When a solid is heated, its electrons acquire higher speeds and strike atoms at the hot surface hard enough to ionize or at least excite them. The energy of motion has been converted to stored electric energy in the atom, temporarily.

On the microscopic scale, an atom emits electromagnetic radiation (light) when one of its electrons undergoes a complex oscillation in the process of changing to a lower state. The emission frequency is the beat frequency between the two frequencies of the before and after states. The emission may be

136

spontaneous or *induced*. The latter represents an increased likelihood of emission due to the presence of similar light waves from nearby atoms. The emitted light is ideally of a single wavelength, but many complications spread it into a narrow band of wavelengths. The light carries away one unit of angular momentum from the atom in the form of photon spin. Although the emission may be viewed as a photon, the photon does not start with zero speed but is created already moving at c. The emission takes a small amount of time to occur, usually much less than a microsecond. The electromagnetic field of the atom vibrates perhaps a million times, depending on the type of atom and the states involved. Each atom emits a single wave train with about this number of cycles and a certain polarization. Other atoms emit similar wavelengths, but there is usually no correlation among the phases and polarizations of the various waves emitted. Thus a collection of atoms emits random light.

Some of the causes of the broadening of spectral lines are doppler shift, uncertainty, fine structure, collisions, and strong electric fields. Hot atoms have fast motion, perhaps kilometers per second. This stretches out emitted light waves if the atoms are moving away from the observer and compresses them if the atoms are moving toward the observer. The result is random small shifts in wavelength. Also, when an atom emits light, it recoils (the photon takes away momentum), and the change in speed doppler-shifts the emitted light.

The time required for emission is not zero. By the uncertainty principle, the lifetime of an excited state is inversely proportional to the accuracy with which the emitted energy can be measured. The short lifetimes of most states (say 10^{-8} second) result in broadened spectral lines (at least a spread of a millionth of an eV, which is substantial in accurate spectroscopy).

Many physical effects, within and outside atoms, give rise to fine structure in the spectral lines. A low-accuracy spectroscope will see simply a broad line instead of a set of fine ones. This is illustrated by the difficulty of seeing the separation between the yellow sodium lines with simple equipment. Collision of atoms at a high rate also affect line width, since the collisions occur while atoms are emitting. And finally, the high voltages and electric fields used to create ionized sources of light spread the energy states randomly and lead to broadening of emitted spectral lines.

12.3 ABSORPTION, FLUORESCENCE, AND PHOSPHORESCENCE

An atom has no objection to emitting the precise wavelengths it is designed for. But *absorption* is a different matter. An atom will absorb only the wavelength that matches the difference between two states. Things are not quite so rigid. The atom will absorb a nearly correct wavelength if something can take up the difference—perhaps the excess energy will go into a little extra speed for the atom. Or the atom happens to be moving away from the light at such a speed that the wavelength is dopplered down to the correct value. Atoms are readily susceptible to radiation being emitted and absorbed by their neighbors because the wavelengths happen to be nearly the correct ones. Thus one atom in a hot gas will absorb light emitted by another atom of the same kind. If the right photon comes along and the atom is excited, it may then spontaneously fall back to a higher state than the original one, emitting less energetic light. This is a *Stokes transition*.

Probability plays a major role in atomic processes such as emission and absorption. No one can say when a given atom will emit its light and drop down from an excited state. But trillions of atoms taken together are quite predictable as to how fast a process occurs. A given atom has a low probability of absorbing a mismatched wavelength. Sometimes it will absorb it, but usually it will not.

Experiment 12-2

If you have a replica grating, examine the spectrum from salt in a flame while simultaneously looking at a background of white light from an incandescent bulb. Look for dark lines in the yellow part of the spectrum. What happens when the bulb is removed? ■

Absorption spectra are the opposite of emission spectra. What is seen is the absence of light at selected wavelengths in perhaps a continuous spectrum. These look like dark lines against a light background. The solar spectrum has many dark lines due to absorption of light by atoms in its upper atmosphere. They are visible against the continuous bright spectrum of the hot sun. Heating salt (which activates the sodium in salt) in front of a white light will give a spectrum of two close dark lines against a yellow background. The dark lines appear exactly where bright sodium lines would appear in other circumstances. Although the hot sodium is emitting as well as absorbing light at certain wavelengths, the light coming from the background source toward the observer is absorbed and re-emitted in many directions, reducing the intensity toward the observer.

When wavelengths unmatched to transitions are shone on an atom, electrons will absorb them temporarily and reradiate them almost immediately with no energy retained by the atom. This is scattering and its efficiency depends on wavelength as Rayleigh showed. The closer the wavelength is to one that is acceptable to the atom, the longer the energy will remain absorbed. Most kinds of atoms have *metastable* states. These are energy states into which the atom can be put but which cannot emit light directly. An atom that has absorbed energy into a metastable state waits until a collision, or another photon of the same energy to pass by. Then it emits the light.

A fluorescent lamp illustrates several processes of emission and absorption. The term *fluorescence* properly refers to the prompt emission of wavelengths different from (and longer than) a wavelength of light that strikes and excites the atom (or another energy input such as collision). In the common lamp, electrons are accelerated to high speed by the voltage applied across the ends. When they strike mercury atoms, electrons are knocked off and/or the atom is raised to a high state, with any excess energy going into motion. Any ionized atoms are accelerated until they collide with more atoms. Electrons meanwhile trickle down through various empty states, emitting various colored lines in the process. Such light would be aggravating to use for seeing, and the light is dangerous to the eye because

mercury emits most of its energy in a strong ultraviolet line (also completely useless for seeing).

Experiment 12-3

If you have a replica grating, examine the spectrum of a fluorescent lamp. If you do not have a replica grating, you can still note the persistence of the short term phosphor in the lamp by keeping your eyes closed while the light is on and opening them the moment you switch the light off. Observe the lamp in a dark room after it is switched off. ■

A phosphor coating is put inside the tube of the fluorescent lamp to absorb the ultraviolet and convert it to a wide range of wavelengths that are visible. Solid materials emit spectral bands (Fig. 12-5A) rather than lines because the electrons interact with each other to spread the possible states into broad bands (Fig. 12-5B). A daylight fluorescent lamp does not have a smooth sunlike phosphor spectrum but rather has extra gases added to produce strong lines of yellow, orange, and red. The optical properties of "solid-state" materials are a vast and growing field in itself. Another property of phosphors is that

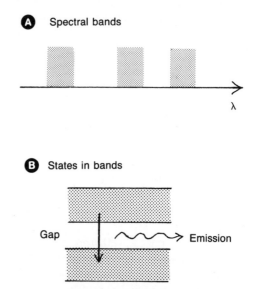

Fig. 12-5. The spectral bands resulting from the many transitions possible between states in one band of states and another in a solid material.

the complex atoms involved store energy in metastable states for long times and emit light gradually. This delayed emission is called *phosphorescence*. Thus the phosphor of the lamp glows steadily even though the 60 Hz alternating voltage causes ultraviolet stimulation intermittently.

Special crayons and chalks, as well as fabrics and printed pictures, can use phosphors. There are also minerals that have one color in normal light and another vivid glowing color in ultraviolet light. If you experiment with so-called "black" lights which emit strong but invisible ultraviolet, do not look at the lamps with unprotected eyes for any length of time. Video screens (cathode ray tubes—CRTs) for television, computer monitors, and oscilloscopes all use phosphors on the inside of a vacuum tube. Electrons in a beam strike the phosphor at controlled places, excite it, and cause it to emit persistent light. Many colors are in use, including special bright red, green, and blue for color television.

Molecules, whether in gases or solid, readily absorb and emit infrared light and even microwaves, in addition to visible light. Some of their electrons can be said to belong to the molecule as a whole and constitute the method of binding it together. Molecules have bands of very closely spaced energy states that result from the interaction of the several slowly vibrating heavy atoms that constitute a molecule (Fig. 12-6). Molecules more readily absorb a

variety of wavelengths because so many states are available between which a transition will match the input energy.

The Raman effect (found by C. V. Raman in 1928) is scattering of light from molecules that results in extra spectral lines besides the wavelength of the incident light. It is not fluorescence, because the resulting spectrum is not the normal spectrum of the molecule. It is not Rayleigh scattering. The effect is weak and usually in the infrared. Light energy absorbed makes the molecule go to a higher state. It then spontaneously emits but returns to a slightly different state either higher or lower than the original. The result is several spectral lines adjacent to the original that provide information about the molecule. Raman scattering is an investigative tool for matter (as most optical work is).

The optical absorption (or absorbance) by a quantity of matter is measured with a *spectrophotometer* (literally a spectrum-dependent light intensity meter). The wavelength used is important, and a curve or graph is obtained that tells how much absorption occurs as a function of wavelength. The amount of absorption depends on the thickness of the material (or its density or concentration in solution). If the sample is very thick, absorption occurs to a certain depth. More precisely, the material has a characteristic depth below which only 37% of the light intensity remains. Going twice this depth re-

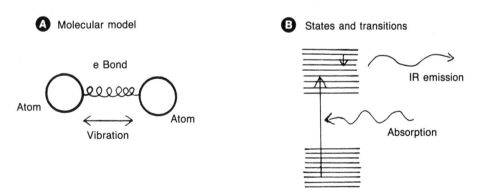

Fig. 12-6. Infrared and visible transitions are possible between the sets of states characteristic of a molecule (modeled with a spring bond).

moves all but another 37%, leaving 14%. This "exponential" absorption continues until the intensity is too low to detect.

When the sample is semitransparent, absorption is measured via transmittance with the spectrophotometer. Selected colors or wavelengths of known intensity are shone through the sample of known thickness, and the amount transmitted is measured. Test wavelengths are obtained with colored filters or more accurately with a *monochromator*. The latter is a spectroscope arranged so that selected wavelengths from a white light source are available as a beam. Colored samples absorb visible light selectively, taking away wavelengths different from the color seen.

12.4 SOURCES OF LIGHT

The generic source of light is atoms. Exotic sources of light such as synchrotrons have been developed in recent decades and produce light directly from moving (really, accelerating) electrons. Thus the ultimate source of light might be said to be accelerating electrons. Many practical types of sources that have atoms and electrons in random motion (heat) are summarized here. Those that extract light from electrons in an orderly way (for example, the laser) are left to the next chapter.

The colors of light emitted from a flame are indicative of its physical processes. With all flames, visible or not, most of the energy is not emitted as light but as infrared or as hot gases. A properly burning flame from almost any hydrocarbon fuel is faint blue in color with narrow spectral bands emitted in the blue and violet near 430, 380, and 350 nanometers from C-H molecules and near 515 nanometers in the green from C-C molecules that are newly formed and excited. The yellow part of a flame appears when the fuel is incompletely oxidized and emits carbon particles. They glow incandescently over the full spectrum, emphasizing yellow although the peak wavelength is deep in the infrared. Figure 12-7 shows the temperature and radiation from a candle flame. A candle cannot have complete combustion because the physical arrangement of fuel, airflow, and flame prevents it. Premixing oxygen with the fuel permits more complete combustion.

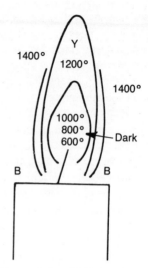

Fig. 12-7. Different light emission regions in a flame (temperatures in °C).

Then the flame has no yellow region but two blue cones, the innermost appearing where carbon dioxide is formed from carbon monoxide. Considering the complexity of the processes in a flame, it is a wonder that it works at all. Flames are still poorly understood, and their stability is therefore remarkable.

Experiment 12-4

Examine the shadow of a candle flame in the sun. This will show where invisible carbon particles are present. ■

Flames are a handy way to obtain the spectra of certain elements that ionize readily at flame temperatures. These include metallic elements with loosely held outer electrons including lithium, sodium, and potassium. The elements can be introduced as salts. The flame melts the crystal and excites the metal constituent to produce several spectral lines. These flame spectra are handy in simple analysis of materials.

High-temperature light sources include about 2000°C in incandescent lamps and about 4000°C in carbon arcs. *Incandescence* refers to the glow of blackbody radiation that occurs because of high temperature. Lamp filaments are commonly tungsten operated sufficiently below vaporization temperature to have a lifetime suited to the application. Anyone who has reduced the brightness of a lamp with a dimmer has noted the change in color and therefore peak wavelength from yellowish to deep red. At low power, the lamp will feel hot without glowing visibly, just as hot irons emit infrared but are invisible, even in the dark. Filament temperature and therefore color is strongly dependent on operating power or voltage. A 10% decrease in voltage, or alternatively, a filament designed for 10% higher voltage than normal, will last much longer as well as give a more yellow light.

The most intense broad spectrum light is obtained from arcs. Two carbon rods with voltage applied are brought to touching, then pulled apart. The air between, and the vaporized carbon, glow intensely as a torrent of electrons passes from one rod to the other. Arcs between metal rods go as high as 7000°C. Arcs emit more dangerous ultraviolet as the temperature rises.

Optical work is often done with the spectral lines from mercury and sodium. These are available from high pressure arc lamps, but the lines are broadened. Electric sparks through air or through gases and vaporized solids cause emission of spectral lines. Because the spectrum of an ionized atom is different from an un-ionized one, this highly energetic way of exciting atoms is needed to see their spectra. If possible, the electrodes between which the spark occurs should be made of the element of interest.

Gas discharge tubes contain small amounts of gaseous elements such as hydrogen, oxygen, neon, or helium, or elements that will vaporize, such as mercury. They are made and operated similarly to commercial neon signs. High voltage causes ionization of the gas as in the fluorescent lamp. If various colors are the objective, rather than examination of pure spectra, mercury is added to neon, and colored glass tubing is used that filters out some of the mercury lines. Mercury provides a line of each basic color—red, orange, yellow, green, blue, and violet, more or less. The noble gases—helium, neon, argon, xenon, and krypton—exist as single atoms in gaseous form. Their spectra as elements are read-ily obtained. More common gases such as hydrogen, oxygen, and nitrogen occur as two-atom molecules. The molecular form has a different and more complex spectrum that tells little about the behavior of single atoms. Ionizing a molecule removes an electron but does not necessarily break the molecule apart. Complex sources are able to separate single atoms from ionized molecules and produce just the atomic spectra. The red and blue lines from atomic hydrogen and a yellow one (not the doublet) from sodium are most commonly used for optical system testing.

Semiconductors include the better known electronic materials silicon and gallium arsenide, but they can also include various sulfides, tellurides, phosphides, oxides, antimonides, and more. These have their electrons in a valence band of states, with empty conduction states above an empty gap (recall Fig. 12-5B). Optically useful semiconductors have the energy difference across the gap equivalent to visible light energies. They absorb visible photons that promote electrons to conduction states, or they emit light as conduction electrons drop to valence states. Impurities cause a sprinkling of extra states within the gap and allow finer control of emission and absorption frequencies. The flow of electric current produces *electroluminescence* in selected semiconductors. Two semiconductors connected at a p-n junction form a light-emitting diode (LED), which emits red or possibly green light when current flows. Semiconductor light sources use low voltage and power and have widespread use as indicator lights and numerical display devices.

Light is generated in a number of uncommon processes. *Thermoluminescence* is the light produced when crystals are heated. It is of archeological value for dating back 300,000 years. The longer a rock or pot shard has been exposed to natural radioactivity since it was last heated to 400°C, the more defects are stored in its structure. Light in proportion to its defects and therefore age is emitted when the material is next heated. Alternatively, the method can give the heating history of the material.

Triboluminescence (from the Greek for rubbing) is light produced by friction within some crystals as they break up and interact with the nitrogen in air. The color is blue to violet. Such light is also produced by friction in adhesives. If other gases are substituted, different colors are obtained.

Experiment 12-4

To see triboluminescence, unroll almost any tape in the dark and watch the adhesive. Also look in a mirror in the dark while chewing hard sugar candy. ∎

Chemoluminescence is "cold" light produced in certain complex chemical reactions. Most chemical reactions either produce or absorb heat. But these emit most of the energy left after the reaction in the form of visible light of a cool greenish blue color. Emergency lights have been made in chemical form.

A *scintillator* is a crystal such as sodium iodide or other material such as plastic which absorbs high energy radiation or particles and emits energetic visible light to be detected by a photomultiplier. Scintillators are needed for detectors in nuclear and particle physics measurements.

Light absorption also has special applications. *Photoconductivity* is the effect of light on the electrical resistance of a material. This is a simple way to make a photodetector. Light assists electrons in achieving conduction states, so that resistance falls when the substance is illuminated. The response is just fast enough to be used in video cameras. *Photochromism* refers to a change of color or opacity by materials sensitive to light. The effect is exploited in photochromic sunglasses which darken (not strictly a color change) in sunlight and absorb rather than transmit light. Certain dyes which change chemical state when illuminated are *photosensitive*. When coupled as sensitizers to silver halide, they are useful in color film.

13

Lasers

Sufficient background has been provided to present you the theory and construction of lasers. A new type of laser is developed every year or so, and it is impossible to cover all types or to anticipate future types. Increasing effort is expended to develop lasers far outside the visible range—at shorter wavelengths than ultraviolet. The laser has become the backbone and principle instrument of optics as well as serving a wide assortment of other sciences and technologies. Its value comes from its high intensity, narrowness of beam, purity of wavelength, coherence, and other virtues. Because the laser is reminiscent of science fiction stories, it is almost unbelievable that such a device could be created. This chapter shows you how the laser brings new dimensions to optics.

No home experiments are offered here, but if you have access to a laser, hundreds of simple yet fascinating experiments in optics are possible. Experiment guides are usually supplied with lasers.

13.1 BACKGROUND FOR THE LASER

The history of the laser, at least in regard to patents, is in some dispute. Its origins are traceable to the maser, a term adapted from the acronym for microwave amplification by stimulated emission of radiation. The name of the laser was created by substituting light for microwave. The principles of the maser and laser are almost the same, and are described by their long names (once one knows the terms). Although the name laser originated for visible light, the term is now in use for devices that work like the laser at almost any wavelength.

After Charles Townes (and independently Alexandr Prokhorov and Nikolai Basov) developed the maser in 1954, he and A. H. Schawlow went on to apply the principle to light in 1958. Another contributor seems to have been Gordon Gould, who recently won his case in the Court of Patent Appeals. The story of the origins of the laser may be murkier than hitherto realized, and additional confusion occurred in the early 1960s because of attempts to classify the work for military purposes. Theodore Maiman constructed the first working laser in 1960 using a ruby crystal. Ruby is aluminum oxide (corundum), with a trace of chromium oxide that gives it red color. Only the chromium ions participate in lasing. Flat polished ends were put on the ruby, with one end highly silvered and the other partly silvered. Energy was put into the device by a gas discharge tube which shone bright light on the ruby, and the laser emission was pulsed (intermittent).

Until a laser was actually demonstrated, many sponsors of research believed that the ruby and other lasers were impossible and cut off funding. Many early ideas did not work, although every later approach seemed to work if done carefully. The first gas laser was developed in the next year by Ali Javan, W. R. Bennett, and D. R. Herriott. Originally made to work in infrared, this helium-neon laser has since become the workhorse of science and technology with its piercing red light. Semiconductor lasers were developed in 1962 and have become the centerpiece of optical communication. Chemical lasers were discovered in 1964. Holography, a spectacular application of the laser often thought to have originated with it, actually was first tried in 1947 by Dennis Gabor using more conventional laboratory light sources.

13.2 THEORETICAL PRINCIPLES OF A LASER

The principles involved in a laser constitute a partial summary of modern physics: metastable states, fluorescence, resonance, stimulated emission, coherence, modes, interferometer, dielectric mirrors, and more. Polarization plays a role. And new concepts for preparing atoms or molecules to ''lase'' are needed (optical pumping and population inversion).

Any source of light, including the laser, produces the light by causing excited atoms to drop to lower states, emitting light. The states of interest are metastable—ones that the atom would stay in for long times if there were not outside influences such as collisions or certain light waves passing by. Just how the atoms or molecules are put into metastable states in the working medium will be deferred for the moment. Once most molecules are in metastable states, the population of states is said to be *inverted* (Fig. 13-1). Normally most molecules

are in ground states, so having most molecules excited is unusual and unstable.

Sometimes an electron in a metastable state will spontaneously jump to a lower state, usually the ground state. It emits radiation of a wavelength that is strongly influential on other molecules similarly excited. The light wave stimulates other molecules to emit the same light and drop to ground states. A special feature of this stimulated radiation is that every time additional light is picked up from another molecule joining the process, the phase, direction, and polarization are exactly the same. The light intensity increases rapidly as more molecules down the line contribute (Fig. 13-2). It is important that most molecules start in the metastable state. Otherwise the light will be absorbed by molecules in the ground state as they jump to the metastable state.

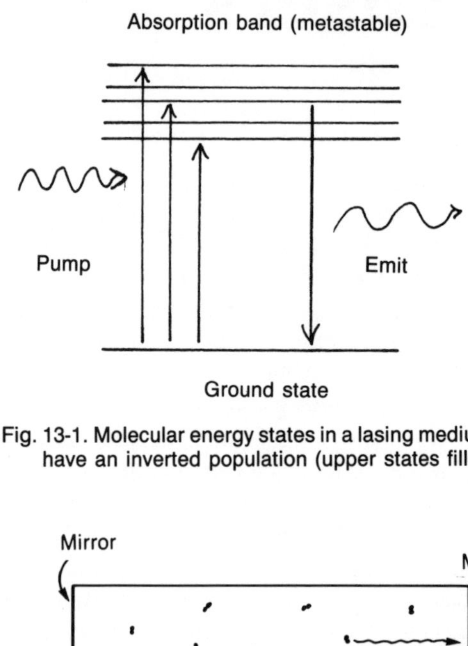

Absorption band (metastable)

Pump

Emit

Ground state

Fig. 13-1. Molecular energy states in a lasing medium must have an inverted population (upper states filled).

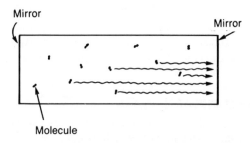

Mirror

Mirror

Molecule

Fig. 13-2. Build up of stimulated emission of radiation in a lasing medium. (A low density of molecules is shown for clarity.)

To build up the light intensity and wring as much energy as possible from the lasing medium, the light needs to make several passes through the medium. Reflectors at each end of a tube containing the medium bounce the light back and forth as in a Fabry-Perot etalon. Typically only a small amount of energy is added to the light on each pass, so that many passes are needed to collect all available stored energy. The reflectors need not be adjusted exactly in spacing, but a half-integral number of wavelengths must fit in the *cavity* between the mirrors. The small size of a wavelength as compared to the tube or cavity length assures that small variations in wavelength will permit a fit. But the mirrors, if flat, must be quite parallel, or plane waves reflecting from them will eventually strike the sides of the tube. When the laser has a resonant cavity, then the light wave is a standing wave in the cavity, being perfectly in phase everywhere in the tube. Other descriptors of the build-up of light energy are *chain reaction* and *avalanche.* As its name states, the laser amplifies light by stimulating emission.

The laser energy must be extracted as a useful beam, so one mirror has a hole or is partially silvered. Intensity is sacrificed, or alternatively, the time to build up intensity is lengthened, so that a useful beam is obtained. The natural diffraction of light limits the angular size of the beam. It must spread at a small angle determined by the wavelength of light divided by the diameter of the exit aperture. A one-millimeter diameter beam will spread at about 0.04°, or grow to 0.6 meter wide in one kilometer. In small lasers, the beam is very small, less than a millimeter in diameter, yet the light is so bright it can injure the eye. Laser efficiencies at low power are usually less than 1%. A common 1 mW laser (optical power) beam comes from about 0.1 W of resonating optical energy, which in turn requires about 10 W of input energy.

13.3 COHERENCE, MODES, AND POLARIZATION

More thorough discussion of coherence is needed at this point, because the laser is an eminent producer of coherent light. Many applications depend on excellent coherence. Atoms emitting light randomly in a thermal source emit waves independently so that there is little relation of phase, direction, and polarization. The wavelength spread is often inconveniently large in an atomic spectral line. The length of a wave train is usually much less than a meter long, not sufficiently long to be used in precision interference experiments. The laser causes atoms or molecules to emit light together, and the resulting wave is coherent for kilometers. The laser beam has its electric field completely in phase. This doubles the time-averaged intensity received by a detector over that of random light.

The variation in wavelength (or frequency) of a spectral line is related to the *coherence time.* The line cannot have zero bandwidth in a practical situation. The coherence time Δt is that time over which one can expect the phase not to change. The variation in frequency Δf is the reciprocal of the coherence time ($\Delta f = 1/\Delta t$). White light is the worst case. With frequencies spread from $4(10)^{14}$ to $7(10)^{14}$ Hz, the coherence time is less than 10^{-14} second. The wave trains average only about two wavelengths at best.

The *coherence length* is the distance over which coherence holds in time and is found by multiplying the coherence time by the speed of light. Coherence length ΔL is equal to the spread in wavelength $\Delta \lambda$. Spectral lines are coherent over a meter or less. Laser beams are coherent over kilometers. The practical benefit is that experiments up to this size can be done, and interferometric methods work well. Spatial coherence is guaranteed from a point source, but may be possible from an extended source. A laser emits almost plane waves, and they are spatially coherent.

When light is partially coherent, interference fringes are visible, but the contrast is much less than for completely coherent light. Instead of intensity varying between zero and maximum, it varies between maximum and a nonzero minimum. There is a background of random light.

Different numbers of wavelengths can fit between the end mirrors of a laser. For a wave to reinforce itself and resonate, it must exactly repeat itself after a round trip of two reflections between mirrors. Each wavelength for which a large exact number fit into twice the distance between mirrors is a harmonic. A large number of harmonics are possible, each differing from the next by a small difference in wavelength that depends on the length of the cavity. The wavelength spread or *bandwidth* $\Delta\lambda$ works out to be $\lambda^2/2L$, where L is the length between the laser reflectors. The frequency differences Δf are found from the speed of light in the cavity divided by twice its length ($\Delta f = c/2L$). It is not very likely that a lasing medium will have several emission lines such that more than one exactly resonates in a given cavity.

Even at just one wavelength, there are a number of ways (modes) that the resonating light waves can be oriented in the cavity. Several modes can exist simultaneously in the cavity. These are arranged like bundles of rods of various shapes and sizes, where each rod represents a resonating wave. The simplest mode is called TEM_{00}, where TEM denotes transverse electric and magnetic modes, and the subscript 00 denotes that there is no subdivision of the resonance in either direction across the cross section of the beam. This mode is shown in Fig. 13-3. A graph of intensity across the beam shows a smooth hump called a Gaussian curve. Intensity rises rapidly from zero at the edges and reaches a maximum in the center.

Figure 13-3 also indicates a few of the other simple modes that might appear when a laser beam shines on a screen. Modes where the beam is divided into resonating sections indicate that the beam as a whole is not coherent in phase. Each mode has its own polarization. Circular modes are also possible in which the beam consists of a set of nested cylinders and its cross section consists of rings.

The spectral purity of the laser beam (the spread in wavelength $\Delta\lambda$ about the center wavelength) is good and depends primarily on the motion of the molecules that lase. These emit randomly doppler-shifted light waves as they dash about in the medium. Ordinary off-the-shelf gas lasers can achieve 10^5 Hz in frequency bandwidth, a frequency stability of one part in 10^9. Bandwidths of 1 Hz are possible with special techniques.

13.4 PRACTICAL CONSTRUCTION OF A LASER

Some improvements in common optical components are needed to make the laser work. For example, the efficiency of gas lasers is as low as 1% gain on each reflected pass through the gas. But more energy than this is lost at each reflection from common mirrors because it leaks through the silver or aluminum coating or is absorbed. Mirrors had to be developed that reflect much better than 99%. Multiple films are employed, designed to maximize reflection rather than minimize it. This works very well at one wavelength, and the laser is usually designed for one wavelength. Another advantage is that other wavelengths that the medium could emit are suppressed because there is inadequate resonance between the mirrors at other wavelengths.

Since planar mirrors are difficult to align to sufficient parallelism and alignment is affected by temperature changes, a focusing mirror at one end or both ends is common. Concave mirrors cause some loss of light gain but make up for it in stability of gain. It is difficult to include mirrors within the tube, so they are usually placed outside. Then the 4% reflection from the ends of the glass tube is of major concern. This problem is solved by using Brewster windows. As shown in Fig. 13-4, these take advantage of the fact that one polarization of light cannot

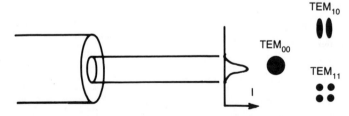

Fig. 13-3. Cross section of the simplest laser beam (TEM_{00}) and some other modes.

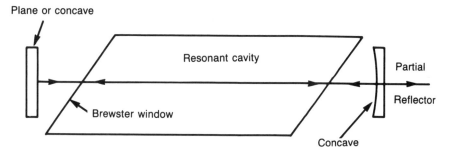

Fig. 13-4. Special windows and reflectors are needed to form an efficient resonant cavity for a laser.

be reflected from a surface at Brewster's angle but must be transmitted at very close to 100% (depending on precision of alignments). Using Brewster windows automatically provides a polarized laser beam, polarized in that direction which can be transmitted out the tube. Those modes with different polarizations are automatically eliminated. To use the laser beam, part can be extracted by a small reduction of coating at one mirror or omission of coating from a small spot.

Tunable lasers are much desired but difficult to achieve, especially in the visible range. Since much important work is done with lasers in the infrared, the availability of closely spaced energy levels in molecules for working in that range is fortuitous. *Raman scattering* is an additional source of usable wavelengths. One wavelength is selected in a tunable laser by using a diffraction grating or other wavelength selective device in the cavity.

Energy is put into the laser by *optical pumping* or similar means. This can be accomplished by wrapping a xenon flashtube around the lasing tube and bathing the lasing medium in high intensity ultraviolet light. Or by using an electrical discharge in the tube to excite either the lasing gas or a helper gas. Or by continuously replenishing liquid or gaseous molecules that react violently upon contact. Or by heating. Pumping must be done continuously if a continuous beam is wanted. Higher-power lasers are often needed in a pulsed mode only. The peak power can be much higher than the average power of a laser. For example, if a pulse of 1 picosecond (10^{-12} second) is wanted at the rate of 1,000 times per second, then a laser with an average power of 1 W can put one gigawatt (10^9 watts) into each

pulse. The total energy put into the material is low, but the instantaneous electric field and intensity is very high. The highest peak powers are obtained by accumulating energy as long as necessary while resonance is suppressed (perhaps by closing a Kerr cell). When resonance is restored, all the energy is released in one burst.

The intensity from a laser is high because all the energy is put into a beam as small as 1 millimeter. For a 1 mW laser, the resulting intensity is 1000 watts per square meter, the intensity of sunlight. This is not the limit. A simple lens can focus the beam to about 10 micro-meters. The intensity increases with the square of the decrease in spot diameter and rises to 10^7 w/m². If laser light were not coherent, a lens could not increase its intensity. The theoretical limit (limited by diffraction) is a spot 10^{-10} meter, the size of an atom. If the beam is wanted large to bathe large areas in coherent light, then a diverging lens can expand it, and a converging lens can restore parallelism. Or the laser beam can be input to the small end of a telescope. Good lens systems are needed for high accuracy, although the light is only one color. At high laser powers, lenses and mirrors must have very low absorption or they will be quickly damaged from heating.

13.5 COMMON LASERS

In the ubiquitous helium-neon gas laser (He-Ne), the input energy pumps helium atoms from ground state to two metastable states. The energy comes from electric discharge in the gas mixture. The helium atoms outnumber neon 10 to 1, and excited helium atoms frequently collide with ground-state neon atoms. The helium energy is usually transferred to the neon because the energy almost matches eight states in neon. The excess goes into energy of motion. The helium atom then returns to its ground state, ready to be pumped again. The neon atoms are in a variety of excited states, so a variety of transitions are possible to lower metastable states, resulting in a variety of wavelengths of light emitted. The ones at 1152 nanometers and 3391 nanometers are in the infrared. A 594 nanometers transition is available, and the famous red light is at 633 nanometers. One wavelength is selected for amplification by designing the resonant cavity for that wavelength. This laser must have a narrow tube for high gain, so the beam must be small.

The carbon dioxide laser delivers very high continuous powers, over a kilowatt at 10.6 micrometers deep in the infrared. This laser is the one usually used to demonstrate burning through steel plates in seconds. It is also tunable over its infrared range and penetrates air well. Being a molecule, carbon dioxide has an abundance of closely spaced energy states due to possible vibrations of the atoms. Nitrogen gas is added to the laser to help excite the carbon dioxide just as in the He-Ne laser. It is much easier to excite carbon dioxide states because the energy involved is much less than visible light (about 0.3 eV per molecule). The carbon dioxide laser is capable of moderate efficiency (15% is very good for any laser) and very high peak power. An optically active shutter is used within the optical cavity to shut down lasing while energy accumulates in the gas. The shutter can be a Kerr or Pockels cell or a mirror rotating rapidly. When the cell is triggered to pass light or the mirror comes into alignment, lasing occurs in a short, high-power pulse over 100 kW. The ratio of peak to average power can be 1000:1, implying a duty cycle of 0.001. A cavity that is triggered into resonance is said to be Q-switched, an old term from electronics.

Ruby was the first and still is an important solid lasing medium. The low concentration of chromium ions in ruby (which give it its color) are the active ingredient. Strong light pumps them into a band of states, from which they relax to a metastable state. That energy is given to the crystal as heat. Then spontaneous and stimulated emission lead to lasing at 694 nanometers (red light). The reflectors are put directly on polished ends of the ruby. Hundreds of other solid crystals and glasses have been made to lase with an assortment of impurities added. YAG and neodymium-doped glass are important examples. Work continues on preventing glass from cracking during high-power use. Many obscure chemical elements have been found useful for lasers as well as in other optical devices.

A series of identical laserlike devices can be arranged so that the first one lases readily and the rest lase only when strong laser light enters them (Fig. 13-5). These amplifiers are prepared with fully inverted populations and help achieve some of the highest powers yet (10^{14} W).

A semiconductor laser is necessarily small and constitutes a special region at the junction of two kinds of semiconductor (the pn junction, Fig. 13-6). Details of semiconductors constitute a vast field of study and must be left to the references. The material is usually gallium arsenide, not the more common silicon. In brief, there are two bands of states separated by an energy gap as shown. Application of voltage drives electrons up to the conduction states and stimulated emission results as they drop to valence states. The energy source is high current at low voltage. The power level is a watt or so and the efficiency is good. When high current pulses such as 10 to 100 A are needed, the pulses must be very short to prevent overheating of the tiny junction. Lasing is in the red or infrared. The use of multiple layers of semiconductor in a heterojunction laser has allowed less overheating and con-

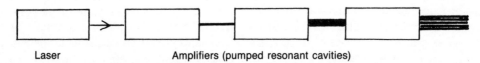

Laser   Amplifiers (pumped resonant cavities)

Fig. 13-5. Amplifying a laser beam with more pumped cavities.

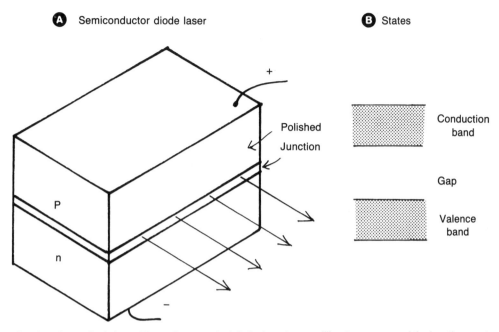

A Semiconductor diode laser   **B** States

Polished
Junction

Conduction
band

Gap

Valence
band

Fig. 13-6. Semiconductor diode laser. The active crystal might be less than a millimeter across, and the junction much smaller.

tinuous operation. These lasers are well-matched to optic fiber communication systems.

Higher power solid state (diode) lasers have recently been achieved by constructing a row of individual tiny solid state lasers in a sheet of aluminum gallium arsenide. Many sheets are then stacked so that one face of the stack consists of thousands of lasers. Continuous power of several watts and power pulses in kilowatts are possible at about 50% efficiency. The lasers all share the same reflecting surfaces, but the output beam is not coherent due to unavoidable frequency and phase variations. Work is proceeding on ways to get many solid state lasers operating in phase. With high power diode lasers available, they can be used to pump up or trigger even higher power, more efficient, narrower bandwidth lasers such as those made from neodymium YAG crystals.

A chemical laser obtains its energy from a chemical reaction instead of electrical or optical pumping. An example is the HF or hydrogen fluoride laser. These two gases react violently when mixed, and the resulting molecules are just what is needed to pump the carbon dioxide also present to its metasta-

ble states. Again, when molecules are involved, the usual lasing wavelengths are in the infrared. Almost any type of energy source might someday be used in a laser. Heat is another common source. Many gases have been made to lase, including all noble gases and gaseous forms of most metals and other elements. Tunable lasers in the visible are often made with organic dyes which inherently have useful transitions at visible wavelengths because they are dyes. The dye is in solution so the laser is liquid. Varying the concentration is one part of tuning. Laser action seems to be easy to achieve once the special conditions are arranged. It does not depend on particular energy levels being available.

The free-electron laser is quite a different approach to producing coherent light that may be tunable over a wide part of the spectrum, beyond infrared and ultraviolet. The basis is a beam of electrons moving at very high speed. When a series of magnets wiggles the electrons in their path, they emit light in phase at a wavelength corresponding to the magnet spacing and the speed. No atomic energy states are involved at all. Cavity resonance is a likely requirement, however. There are many

149

practical difficulties, so these lasers are in their infancy. High efficiency, tunability, and high intensity are the advantages possible.

13.6 SOME APPLICATIONS OF THE LASER—SPECIFICALLY OPTICAL

Upon observing a laser beam land on any nonreflecting surface like paper, one notes immediately that the light seems to dance and dazzle, and the surface appears speckled. This is not just a visual reaction to its purity and brightness but an interference of diffusely reflected plane waves. Speckles can appear directly as a result of interference from the rough surface and on the retina. When the eye is moved, the speckles move differently, depending on whether one is nearsighted or farsighted!

The usual optical experiments can be done easily with a laser. Sending the beam through one slit can produce diffraction fringes centimeters apart on a distant wall. Using a tiny round hole produces a series of circular fringes. A diffraction grating provides about three orders of dispersion, with three spots spread across the room. (The wavelength spread of the laser is too small to resolve with an ordinary diffraction grating.) Lens arrangements can be demonstrated and tested. Multiple reflection occurs between almost any two parallel surfaces at hand, resulting in multiple beams. Scattering from small particles suspended in air or in water is impressive. Beam polarization is easily detected. A Michelson interferometer is readily set up on a stable table.

An unusual interferometer is the ring laser, a specialized Sagnac interferometer. It is a gyroscope sensitive to changes in orientation. Instead of rotating parts, a laser beam is split to go in two directions around a closed path, triangular or square, bouncing from mirror to mirror (Fig. 13-7). The laser can be an outside source or be part or all of the path segments. In effect the optical circuit is a resonator without ends. Waves going in both directions interfere until there is a standing wave. The mirrors must be very good because the beams have many opportunities for loss during many reflections. Reflection of 99.995% has been achieved, and the number of circuits a photon makes is basically limited by how many times it is reflected before lost.

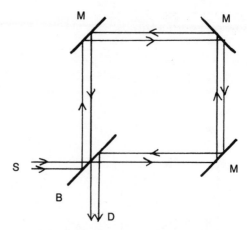

Fig. 13-7. A ring laser (Sagnac interferometer) is very sensitive to rotations. (D is a light detector.)

Any rotation of this system produces a difference in travel time between the two contrary paths and the stationary standing wave appears to move. The passing fringes can be counted just as in an interferometer and tell the rate of rotation. Some subtle difficulties had to be solved before practical optical gyroscopes were ready for use on jet airplanes. The magnitude of the effect can be increased by using a coil of optical fiber about a kilometer long, although this approach is still being perfected.

An uncertainty principle holds for light in regard to its amplitude and phase. One cannot know simultaneously both the amplitude and phase of a light wave to high accuracy. The most steady perfect laser beam has variation in the number of photons traveling. Reducing random fluctuations in amplitude is one way of increasing accuracy, and this results in more variation in the phase. Conversely, reducing phase variations or jitter results in more noise in the amplitude. Accomplishing this involves two new procedures called "squeezing light" and four-wave mixing and makes use of nonlinear media (discussed in the next chapter and the article by Slusher and Yurke). Besides confirming theory, squeezed light with lower noise should be useful in precision interferometry and many other applications.

An unusual quantum phenomenon that will go unexplained here is the photon echo. When two laser pulses enter a ruby crystal, three emerge. The crystal somehow creates a duplicate pulse.

For artistic fun, the laser is employed in light shows. Tools of the trade include: smoke to make beams visible, crossed diffraction gratings, Ronchi rulings, symmetrical arrangements of pieces of diffraction grating, specially designed diffraction elements, motor-rotated mirrors, mirrors mounted on speakers and vibrated by music, polarizers, crumpled plastic, lumps of dry plastic glue, and the like. Diffraction gratings should be coarse so that many orders are visible and bright. A Ronchi ruling or filter is a very coarse grating consisting of opaque and transparent strips. When two gratings are crossed, the row of diffracted laser spots becomes a grid of spots in two dimensions. If the crossing is done at an angle other than 90° and/or several gratings are so combined, the pattern of spots is more complex. In the references are books and articles on making complex diffraction elements (also called optical transforms). The laser beam must have substantial power (perhaps 10 mW) for large displays and therefore is dangerous to eyes. Federal safety laws govern public laser shows. Parts of the beam can be tapped off by reflection from small glass sheets and sent to various display stations. If the laser has several colors (and is expensive), the colors can be separated by a prism for use.

13.7 HOLOGRAPHY

A popular laser application is *holography*. As the Greek root in the name implies, a wholeness of a scene is photographed using laser light. Coherent plane waves illuminate an object (Fig. 13-8A) and are reflected in many directions from each point of the object. Some reach photographic film. Meanwhile the same plane waves (a *reference* beam) are directed to the film by a mirror without encountering any other object. The arrangement is not critical. At the film, the two sets of waves interfere in a complex way that contains information about the position of every point on the object. When Gabor developed holography, only conventional light sources with very short coherence lengths were available, and experiments were very small. The first laser holograms were done by Emmett Leith and Juris Upatnieks in 1963. These were large, and more free of problems because the reference beam could be separated from the illuminating beam.

Film for holography must be extremely fine grained, because the interference pattern on it has structure as small as the wavelength of light. When developed as a transparent plate, the *hologram* shows no detail except traces of complex fringes at an extremely small scale. Dirt in the system may produce Fresnel rings and other patterns that have nothing to do with the image wanted. But when plane coherent light of any color is shone through the plate, the eye perceives a three-dimensional virtual image of the original object (Fig. 13-8B). It is visible in the same direction as occurred in the original setup as if the eye were looking along the direction of the

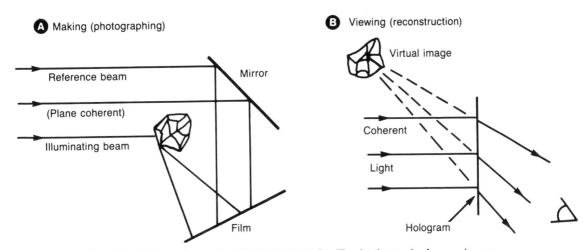

Fig. 13-8. Holography as done in two stages using illuminating and reference beams.

151

original laser beam. Moreover, a real image is produced in space, invisible unless fine particles are present to scatter the focused light. The real (conjugate) image is reversed in the sense that distant points are close and vice versa.

Producing holographic images is called *reconstruction*. There are few limits to the arrangement of incident light, object, and film. The setup shown here conveniently separates the images for the viewer. No lenses are needed for any part of the holographic process. The principle difference from conventional photography is that the amplitude (not just the intensity) and the phase of light from every point of the object are used. It is sufficient that the illuminating intensity be no more than ten times larger than the reference intensity.

The color of the image is the same as the viewing light. Not only is depth perceived, but as the eye is moved, different sides and perspectives of the object are seen. One cannot see behind the object because no laser illuminated it there. Projecting the image as a real one on a surface is infeasible because no surface can show a three-dimensional image. Science fiction envisions real images in midair, and this has been done experimentally. The problem of creating an image in full color is partly solved. Three colored lasers such as red, blue, and green are used to make one hologram in thick film. Reconstruction can use white light. Holograms have been made using other types of coherent waves (sound, microwaves, x-rays, etc.) and then viewed with laser light.

Holographic plates can store many images. Thick film and transparent light-sensitive crystals such as lithium niobate can store the interference patterns in three dimensions. Hundreds of different holograms can be piled into one plate and retrieved according to illumination direction or color. When laser light of the proper direction or color is provided, the selected image is seen as a result of diffraction from the appropriate interference pattern in the film. The light seems to "know" which set of interference fringes embedded in the film belongs to a particular image, as no other set produces a pattern that focuses to form an image. The pictorial information storage is highly ordered despite the seeming randomness of the actual storage elements. The information is truly stored everywhere at once in the hologram as its name implies. If half the hologram is cut away and discarded, the whole image is still there. Even a small corner of the hologram can produce the complete subject, but rather fuzzily. As the size of the hologram shrinks, the number of light rays reinforcing each point of the image is reduced and the pictorial noise level rises.

Making the hologram requires coherent planar light and a stationary subject. But various ways have been found to view the hologram in white light. For example, a filter is sold that converts slide projector light to quasi-monochromatic, and in this a hologram can be viewed clearly. Without a restriction on the bandwidth of the light, the image will be smeared in space and in color as each wavelength in white light produces an image in a different place. A spectral source (monochromatic) is better and does not require coherence. A point source of white light works for specially prepared holograms, rendering the image in the colors of the rainbow. The method was invented by Stephen Benton in 1969, and the explanation is complex (see the references). Briefly, a second hologram is made using light through a slit and the first hologram. When it is viewed with white light, the image is seen in one color at a location where the image of the slit would be. Moving the eye up or down brings another slit image into place, and the image of the object has a different color. The coloring will vary through the full spectrum, and the image is quite bright. Moving the eye away causes the whole image to be colored like a rainbow. The use of the slit has eliminated vertical parallax, but two eyes in a horizontal plane only need horizontal parallax to see depth.

More ambitiously, the composite or multiplex hologram combines moving scenes in one hologram. Conventional photography is used to take the subject at different stages of motion from different directions. The multiple images are then converted to narrow strip holograms and combined on one plate by using a vertical slit. Then as one moves around the hologram set up for viewing, the subject is seen to move. Each eye sees a different hologram and therefore sees depth due to horizontal parallax. The photography is easy, dispensing with the long exposure times or intense laser sources needed to record the initial holograms. The hologram can be made for white light viewing.

Theoretical aspects of holography are many. The sum of two phasors for every point of the object arrives at the film, which records the sum of the intensities of direct and reflected waves and an additive or subtractive amount of intensity that results from any phase difference. If the film has depth, a pattern is recorded throughout. Two sets of plane waves (denoting no object) result in exposure of the film in the form of closely spaced sheets of silver grains at some angle throughout its depth. The hologram of a point is a Fresnel zone plate. To understand reconstruction, recall that plane waves passing through a zone plate converge on one point, the focus. They also seem to be diverging from another conjugate focus. This process is repeated for every point of the object to form the image. The hologram is a dense mixture of zone plates. The outer zones become finer and finer. When they cannot be resolved with the film grain, they go unrecorded.

Practical holography requires fine grain film and long exposures or a bright source. The fringes may be spaced a wavelength apart. Film is available with grain this fine (made for spectrographic work), and its ASA speed is about 0.03. The objects must hold still for seconds or even minutes. Laser light is not very intense when spread over a large object. The hologram can be made nearly transparent by bleaching the film. No matter whether the film is copied positive or negative, the same hologram results. Larger-grained film can be used in special cases where the reference waves can be spherical to accompany the spherical waves emitted by object points. The resulting hologram is called a Fourier transform hologram.

As will be seen in the next section, many uses have been found for the holographic process, but there are few widespread applications of holography just for image-making. Holograms of objects are widely available as curiosities and artwork but are not in serious use in industry or entertainment despite early predictions. The possibility of three-dimensional images, colored or not, surely has ap-plications, but the technology has been challenging and expensive. Recent computer-calculated holograms have been displayed that show the use of models for those industries needing visual models—for example, architectural design, medical study of the internal living organs, and product design.

13.8 FURTHER APPLICATIONS OF LASERS

Interferometric uses of laser light in astronomy are to measure the distance from Earth to the moon to a precision of less than 10 centimeters, to better establish the orbit of the moon, and to measure the size of the Earth. A retroreflector was placed on the moon by Apollo 11 astronauts in 1969, and astronomical telescopes are used to send ruby laser pulses to it and receive the return pulses about three seconds later. The return pulses are spread to almost 20 kilometers in width by diffraction and barely distinguishable from optical noise. But timing is very accurate, and benefits of this work include measuring continental drift and locating the Earth's poles more accurately.

More locally, laser beams provide precise straight lines for surveying and locating objects on the Earth's surface. Monitors of fault movements for earthquake prediction and other small movements of the Earth's crust use lasers in long interferometers. Sometimes the beam path is in an evacuated tube, sometimes in air. Lasers in two colors are used to correct for atmospheric effects along the beam paths. Changes in area of land are detected by a triangle of laser beams. Large irrigated farms are made level by grading the land according to a reference laser beam. Physicists are also studying how certain laser wavelengths can be used for a new definition of the standard of length, the meter.

Laser light shone through or on stressed or vibrating materials reveals the effects of the stress or patterns of vibration. Defects can be found nondestructively, revealed by their abnormal vibrations. Although moving objects cannot be holographed clearly, the moments when their motion corresponds to half wavelengths can. Objects that are supposed to be duplicate can be compared with a hologram to locate small differences. Time-lapse interferometry compares the pattern in a medium without disturbance with the pattern induced by a disturbance—for example, the turbulence behind a high-speed object. Laser light shone through the atmosphere and received by detectors can reveal the amount and structure of turbulence in the air.

Holography can be done with other coherent waves such as sound waves of any chosen frequency in acoustical holography. The hologram from sound waves can be produced on a liquid surface (see the article by Metherell)—but the hologram can be viewed in laser light if it consists of dark and light patches corresponding to acoustic interference fringes.

Remote sensing by laser includes identifying unknown substances at a distance (Fig. 13-9) and determining position and speed of distant objects and windspeed. The process has been called lidar, by analogy to radar. The distance can be tens of kilometers, and remote temperatures and pressures can be measured as well. The scattering and return of small amounts of laser light is caused by particle (Mie) scattering, Rayleigh scattering, and Raman scattering. Clouds of volcanic ash, air pollution, and ordinary rain and snow have been tracked and analyzed with the returned light.

Solids and liquids have been very difficult to analyze at a distance because of their complex molecular responses, but gases are easily identified, especially many common pollutants in the air. There is a lack of lasers in the infrared band where many organic substances have spectra. Induced fluorescence or *fluorimetry* is the excitation of unknown substances with light, followed by analysis of emitted spectra. For example, oil slicks can be identi-

fied from the air with an ultraviolet laser. Detecting the density and temperature of water molecules with three laser beams allows remote sensing of temperature and humidity in the atmosphere. In a different realm, laser-induced fluorescence in tiny ruby crystals in high-pressure experiments provides a measure of the pressure. And water depth can be measured with the scatter from pulsed blue-green light aimed straight down.

Raman scattering, resulting in side spectral lines when strong light is incident upon a molecule, is much increased when powerful laser light is used, tuned to a spectral line. The molecule is stimulated, and coherent scattered light emerges at wavelengths corresponding to nearby transitions in the molecule. The process is useful both for studying the energy structure of molecules and for generating high intensity coherent beams at a wide variety of wavelengths. Raman and other methods use the laser to study the fine structure of spectral lines of atoms and molecules. Tunable laser light is swept along the spectrum, and absorption lines are noted. Since the light can be polarized, atomic states of different polarization can be distinguished. Many techniques for avoiding doppler shifting had to be developed. The spectrum of hydrogen has been explored in such detail with dye lasers as to allow investigation of the properties of the proton and electron itself.

Raman scattering is also used to monitor pressure changes in high-pressure cells. A sample of any material, such as a mineral expected deep in the earth, is squeezed between two diamond "anvils" in a simple hand-operated press. Pressures over a million atmospheres are now achieved. The dia-

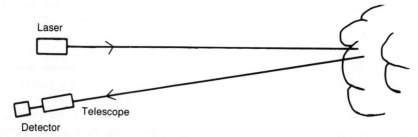

Fig. 13-9. Remote sensing with a laser to identify distant substances, measure temperatures, or other studies.

monds are transparent to a wide band of optical wavelengths, allowing laser monitoring or even laser heating of the sample. Pressure is measured by using laser light to monitor changes in a bit of ruby crystal included in the chamber.

The study of atoms is also helped if they can be stopped. The pressure of concentrated laser light has been used to manipulate and capture single atoms (for example, sodium). Three pairs of laser beams in each of three dimensions intersect in a region called "optical molasses." Atoms can be trapped there about one second, quite a trick for very tiny, agile objects. About 30,000 photons of yellow light were needed to halt a sodium atom in one experiment. Each takes a bit of energy away from the atom as it is absorbed from one direction and re-emitted in a random direction. The changing doppler shift during these interactions makes the experiment very challenging.

The pressure of laser light is being used to move and sort larger particles, including viruses and cells. The shape of the laser beam results in forces that center particles in the beam (when refraction conditions are correct) as they are moved along. When the light is aimed upward, particles can levitate against gravity. The spin carried by light will make the particle spin. A particle can be held in a focused beam as if it were in a tweezers. Very accurate measurements of small particles are possible with laser manipulation. Cells and other particles are sorted when a fluorescing dye incorporated in selected ones is activated by a certain wavelength. The selected cell or particle then emits a longer wavelength and thereby signals its presence.

Extremely short laser pulses help scientists to "see" what happens in the rapid world of molecular reactions. Working like stop-action photography at a time scale of 10^{-15} second, laser pulses allow the behavior of molecules to be determined in between vibrations and collisions. The shortest pulses at this time are a little less than 10 femtoseconds (10^{-14} second) and contain just a few wavelengths. Generating ultrashort pulses involves new arrangements such as passive mode-locking or colliding pulse mode-locking. A short pulse (as will be seen in Chapter 14) requires the generation of many fre-

quencies of light instead of one. The mechanism involving the optical pumping of dyes is left to the references.

To shorten a laser pulse, it can be processed in a cavity, with the leading edge amplified and the trailing edge absorbed. Another method of optical pulse compression uses *chirping*. The latter term comes from radar where pulses are generated with swept frequency. The chirp is then compressed by passing it through a medium where the speed of light decreases with frequency. The lower frequency parts of the pulse catch up with the higher frequency parts and compression results. The chirping of light can be accomplished in an optically active cell (Kerr), and the compression is done in a dispersive medium. Or optical fiber and a pair of diffraction gratings can do this as shown in Fig. 13-10. The limit to short pulses is the uncertainty principle. Shorter pulses require special lasers that produce wider bands of coherent light over many adjacent frequencies. This is quite different from the original goal of a laser—one frequency. Besides studying fast processes in molecules and materials, ultrashort laser pulses aid in the operation of ultrafast electronic circuits. Coupling light to electronics requires the speed of a Pockels cell.

Photochemistry or laser chemistry is done by exposing particular molecules to a carefully selected wavelength that excites them to react. Many reactions are slow or do not occur unless an activation energy is supplied. Simply heating them may cause unwanted reactions, whereas tuning the reaction to one pathway can be done by laser. When only a sample portion of a chemical is activated by laser for analysis, the ongoing reactions in a complex system can be identified and studied. This approach is especially useful for understanding flames and burning in many settings—e.g., candles, rocket engines, internal combustion engines, coal burning. A goal is to find ways to modify combustion for more fuel efficiency and less pollution from byproducts.

Laser chemical analysis attempts to identify elements and compounds and their reaction pathways. The ultimate goal is to identify single atoms and molecules in an unknown material. In multiphoton ionization, intense lasers are used to pump several

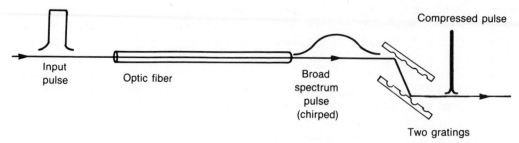

Fig. 13-10. Optical pulse compression.

photons so rapidly into each atom that wavelength selection is irrelevant. The atom is ionized by brute force. Of course, ionization occurs much more readily if the total photon energy supplied is tuned to match the expected atomic transition. Once ionized, the atom can be identified with conventional means that work for very small quantities, including fluorimetry. To obtain a vapor sample for study, another more powerful laser may be used first to blast a bit from the surface.

An *isotope* is an alternate form of a chemical element which has slightly different mass in the nucleus but acts chemically about the same. Until the laser became available, separating the various isotopes of an element for a multitude of uses was extremely difficult and costly, particularly if commercial quantities were needed. The energy levels of different isotopes differ by small amounts such as one part in 10^5, and an accurately tuned laser can discriminate between them. Laser energy is pumped into a chosen isotope, putting it in an excited state or even knocking an electron out. Once ionized, the chosen isotope is easily pulled aside from a beam of atoms. This requires the atoms be in a gaseous state in a vacuum. Often the element is part of a molecule, and exciting one isotope makes the molecule unstable and fall apart. Instead of providing all the energy in one laser beam at one wavelength, it is often necessary to use a two-photon or multiphoton process. One photon excites the isotope, the second ionizes it or dissociates the molecule.

The destruction of materials with intense laser light can occur because of heating or because of high electric fields. The electric field in a light wave can be calculated from its intensity as shown earlier. A field higher than about a hundred million volts per meter tears the outer electrons away from some atoms and produces a hot soup of ions and electrons called a *plasma*. This is optical or dielectric breakdown. Only about a million volts per meter is required to cause it in air, so that intense laser beams in air are made visible like straight lightning. Materials such as crystals and some atoms are more tightly bound so an electric field up to a thousand times greater is needed to disrupt them.

Laser heating may be constructive or destructive. Carbon dioxide lasers are increasingly used in industrial operations at a power up to 100 kW (continuous). Applications include cutting, heat treating, diffusion of hot material, and welding. Big lasers are complex but efficient, involving gas and water cooling systems, waste gas evacuation, and gas replenishing. Major efforts have gone into concentrating powerful (10^{13} W) laser beams from neodymium glass systems on frozen pellets of fusion fuel to demonstrate laser-activated fusion power and study associated plasmas. Nuclear reactions that release neutrons have been achieved.

Lasers are especially useful where the only heating wanted is in a thin surface layer, or where focusing in a transparent material prevents significant heating anywhere but at the focus. Fine low power work is done with glass or YAG lasers pulsed in the 100 W range for drilling small holes in materials such as diamonds, cutting small parts, microwelding, and heat treating. The heating depth is only about 10 nanometers in metals. Lasers provide one way to make ever smaller microelectronic circuits by both physical and chemical action. Melting to anneal the surface of substances such as silicon is stymied when laser pulses are very short (a few femtoseconds). The surface acts as if it melted, but the temperature is not nearly high enough. Ultrashort power pulses seem to open up new technical realms.

The most delicate laser work is medical, mainly cutting and repairing retinas. In experimental biology laser spots as small as the wavelength of light can make lesions in (literally, destroy) selected parts of cells, including the genetic material. One watt of power is sufficient when so concentrated. Medical surgery is being tried, especially when there are benefits from being able to focus the light through transparent tissue to reach a target spot and to control bleeding.

Another constructive use of the destructive power of the laser is in microphotochemistry, the alteration of surfaces in small amounts to produce microcircuits or other fine structures. The working size is about 1 micrometer now. If it is desired to add a controlled amount of an element to a surface, a gas containing the atoms needed is photodissociated locally by selected focused laser light, and the resulting ions react with the surface. Or the laser energy is focused to heat the surface directly in selected locations, leading to reactions with nearby molecules (photolytic method). Focused laser energy can also remove or etch surface material in small regions. If it is not convenient to move the laser beam around, the material itself can be moved. Long narrow strips of metal can be deposited on silicon by these methods. The possibilities for complex and unusual structures on a microscale are yet to be fully explored.

This is only a brief sampling of uses of the laser. Many commercial products such as the bar-code scanner at checkout counters and the laser-read compact disk recordings of music must be omitted. Laser applications are hindered by lack of more widely tunable lasers and lack of far ultraviolet lasers. The many practical problems include damage to lenses and mirrors due to local heating in higher power experiments. More applications of the laser are included in Chapters 14 and 15 on information transmission, processing, and storage and on nonlinear optics. Again, if the reader has access to a laser, the experiments are exciting but can be dangerous. Remember, infrared laser beams are invisible, but their power can temporarily or permanently damage your eyes in a short period of time.

14

Light and Information

Besides the use of the laser to study the behavior of matter, modern optics serves society's information explosion. Because of its small wavelength, light at visible and near-visible wavelengths is ideal for the storage, transmission, and processing of information. Present methods are bulky and inefficient compared to the possibility of manipulating light. Light has long been exploited for communication, from campfire signaling at night to waving lanterns and flags, interrupting beams with shutters, and flashing sunlight with mirrors. All methods required coding of information in an appropriate manner.

This chapter provides, at an introductory level, the theory approach to light and optics. Application is at three stages—the information in a picture or image, the transmission of information from place to place, and the processing of information. Conventional computers are not the only means for processing information. To illustrate the concepts, many applications to present and anticipated technology are mentioned.

14.1 INFORMATION

For modern technology such as computing, transmission, and storage, a rigorous and general definition of information is needed. It should be useful for light as well as for electronics and pertain to all past, present, and foreseeable technology. Information is measured in *bits* (from binary digits used in most computers) and is defined in terms of alternatives. If a certain signal is either present or absent, that is information. If the alternatives are the 26 letters of the English alphabet, the information contained in one letter is that which distinguishes it from 26 alternatives. The fate of nations can be determined by the receipt of one bit of information ("1 if by land, 0 if by sea").

If silence is defined as "0" and the presence of a simple signal as "1," then this is a binary code for information consisting of two kinds of bits, 0 and 1 (Fig. 14-1A). The signal can be a sinusoidal light wave, but if it continues on and on, it cannot provide more than one bit of information. There is no

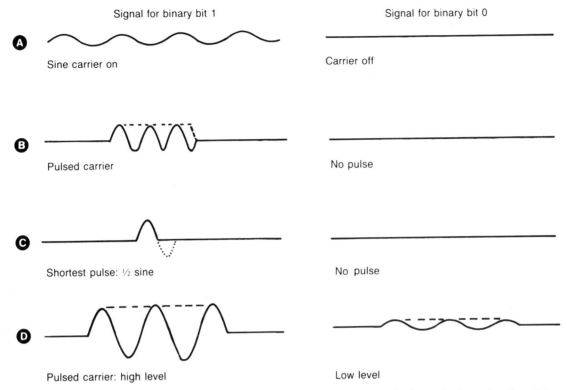

Signal for binary bit 1 Signal for binary bit 0

A Sine carrier on Carrier off

B Pulsed carrier No pulse

C Shortest pulse: ½ sine No pulse

D Pulsed carrier: high level Low level

Fig. 14-1. Binary information carried by waves. The signals are shown with amplitude graphed as a function of time.

variation—no alternative—to compare it with. Such a signal is better used as a carrier of information. To send more variety of information, the carrier must be pulsed or modulated. For example, the pulse shown in Fig. 14-1B consists of turning on and turning off the carrier. This is *pulse code modulation* (PCM). A pulse can last a long time and contain many cycles of a sinusoidal carrier, or it can be shrunk to just one cycle—indeed, one-half sine wave looks just like a pulse (Fig. 14-1C). If the duration of the pulse is a code for the information it represents, then the modulation is said to be *pulse width modulation* (PWM). A short pulse could stand for a dot and a long one a dash, as in the Morse code. Or 10 different lengths of pulse could denote the digits 0 through 9. Another way to have pulse code modulation is to use a different amplitude of carrier for each kind of pulse, as in Fig. 14-1D. Then the recipient of the signal need not be careful about

knowing where the silent parts are so that they may be interpreted as 0s.

A fundamental fact about information emerges. The maximum rate at which information can be sent or processed (measured in bits per second) is determined by the highest frequency carrier that can be used. The maximum pulses per second is double the carrier frequency. If the sinusoidal carrier has frequency f, then the shortest pulse length is one half period or $1/2f$. Sinusoidal waves are basic to the study of information. Everything else, including pulses of any shape, can be expressed in terms of these most simple waves or functions. And they are the simplest carriers to generate. Light is automatically sinusoidal.

A continuous range of frequencies (or wavelengths) is called a *band*. It has a bandwidth as used earlier, expressed as a number in hertz (Hz). This is the range from the lowest to the highest included frequency. If the carrier is one fixed frequency as in, for example, radio transmission, how is it that there is a definite, sometimes large, bandwidth when information is being sent? Consider the simple case of transmitting music with a carrier at 1 MHz by means of amplitude modulation (AM) as graphed on a frequency scale in Fig. 14-2. For present purposes, there is no need to distinguish between information as bits and information in the form of continuous or analog signals like music. The highest frequency one needs to hear in the most precise reproduction of music is about 20,000 Hz (20 kHz). Therefore the carrier must vary between 1 MHz and 1.02 MHz (1 MHz + 20 kHz). For technical reasons, the actual variation is between 0.98 MHz and 1.02 MHz, a bandwidth of 40 kHz. Varying the amplitude of a carrier at a rate up to 20 kHz results in a complex signal that acts as a band of frequencies. (It might be possible to compress the transmitted information digitally. But if every nuance of the most general kind of sound, however complex it might be, is to be transmitted, then 20 kHz is the absolute minimum bandwidth required.)

The band for radio transmission could become crowded quickly. AM radios are assigned frequencies from 0.5 MHz to 1.6 MHz. In that total band of 1.1 Mhz, only about 50 different modulated signals or *channels* can be crowded (more since less than 20 kHz bandwidth is accepted). Very precise tuning is needed, or the signal from an adjacent channel will "leak" into the one selected. Another kind of radio transmission uses frequency modulation (FM), in which the frequency of the carrier is varied in proportion to the frequency of the information signal. For better performance, a bandwidth of 200 kHz is assigned to each FM radio transmitter, although less than half of this is used. The carriers are between 88 MHz and 108 MHz. The minimum bandwidth needed to carry 20 kHz information would be 20 kHz, so FM does not enable more efficient coding.

Human activity needs more channels, and, for television, more channels with more bandwidth. Citizens band buffs might recall the problems getting enough channels (40) into the 0.27 MHz band assigned near 27 MHz. With poor voice modulation, 40 channels at 5 kHz each require 0.2 MHz and very careful tuning. How it is determined will be shown in the next section, but a television (video) signal needs about 5 MHz because pictures are more complex than sound. Clearly a 1 MHz carrier is hopeless, and much higher carriers have been chosen for television. Twelve channels occupy 72 MHz in several bands between 54 MHz and 216 MHz (6 MHz is assigned to each for less crowding). Still higher carriers are needed. Channels 14 through 69 are in two bands from 470 MHz through 806 MHz (UHF). The trend is to higher and higher frequencies.

In the satellite age, telecommunications satellites must handle hundreds of television channels and tens of thousands of voice and computer channels to be economical. Carriers clearly must be up in the gigahertz range (thousands of megahertz). That still does not seem to be sufficient. Clearly, using light frequencies (10^{14} to 10^{15} Hz, or up to 1 petahertz) is an advantage. Modulated infrared (10^{14} Hz) has room for ten million video channels. Light as a carrier of information is being tried in several ways now.

14.2 NOISE AND ENTROPY

The sender or user of information is hampered by noise, or what amounts to almost the same thing, deterioration of the information. A crisp pulse generated at a transmitter (Fig. 14-3) and sent down a channel as a signal arrives as shown, sloppy. The original pulse has a very rapid rise time and ends with a rapid fall. A perfect pulse is not only impossible to create but also unnecessary. A realistic pulse

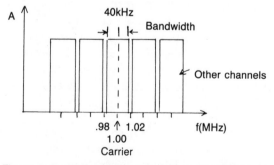

Fig. 14-2. A communication channel has a bandwidth determined by modulation of the carrier and is analogous to a spectral line with width.

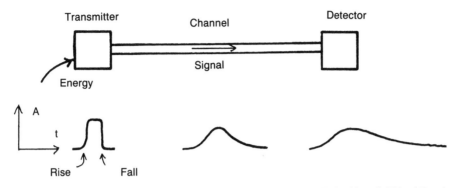

Fig. 14-3. Ideal pulses originating in a transmitter are degraded by the limited bandwidth of the channel and further by the detector.

cannot rise any faster than the sine wave in its carrier. The transmission channel responds slowly to the pulse and stretches out its end, and the detector of the pulse is more sluggish. What is received may be almost indistinguishable from the next pulse. The dead time in between, which may denote a "0," could be imperceptible.

There is more to noise. Nature's materials are very restless, and atoms and molecules are continually emitting radio noise—indeed, all over the spectrum. The result is many small random pulses as shown on a plot of amplitude versus time in Fig.14-4. Up to a practical limit, all frequencies are mixed together in ordinary random or *white* noise (named because white light is a random mixture of all colors). The pulse representing transmitted information may be almost buried in the noise, as shown. Clearly there is a minimum energy that must be put into information in order that it be detectable in a given situation with noisy channels and detectors. Because there is noise all over the spectrum, it does not help to make the band wider in order to send more information. There is a fundamental limit to the information that can be carried in a noisy channel.

The minimum amount of power P (in watts) needed to carry B bits per second can be calculated from a fundamental physical constant $k = 1.38(10)^{-27}$ joules/degree and the temperature T thus:

$$P = 0.693 \ kTB$$

The constant k represents the amount of noise

energy generated by materials at temperature T. It is very small, but difficulties arise when the signal of interest is from distant stars, or bounced off Venus, or sent from spacecraft near Uranus. Terrestrial difficulties with noise versus signal power are also numerous, and using the theoretical minimum power is never sufficient. There is another theoretical limit, too. When the energy in the signal is low enough or the frequency high enough, quantum noise also becomes important. Photons are unpredictable in accord with the uncertainty principle. When n photons are involved in a signal, the minimum variation in photon count is the square root of that number. At the frequency of visible light, the energy in the signal must be at least one tenth of typical excited atomic states in order to be detectable.

There are different efficiencies of coding information to make best use of a channel. Pulse coding

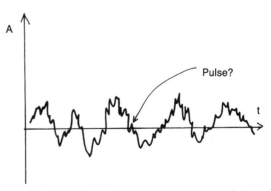

Fig. 14-4. Even pulses can be buried in noise.

avoids noise the best but is wasteful. More information can be crowded in by varying the amplitude at the expense of confusion with noise. The most efficient coding makes the signal look like noise! This is understandable because noise consists of a very complex mixture of signals from atoms. If one had the key to the code for noise, and could work rapidly enough, one could decipher from noise enormous amounts of information. The information in a hologram is nearly indistinguishable from noise and therefore very efficiently stored.

Amplitude modulation is not efficient when a signal is speech or music because most sounds consist of tones held for relatively long times. If the music were very avant-garde and consisted mostly of random noises, nearly the full theoretical channel would be needed to transmit it. Information that is nearly indistinguishable from noise is called *chaos*—a term adapted to the new complexities of the information age. The reference by Crutchfield on chaos shows a picture scrambled carefully many times to form a bland mess, then scrambled the same way many more times to result again in the original picture. The results of very complex coding are orderly nevertheless.

Why is binary coding used, if it is so inefficient? Its weakness is its strength. If at any time in the signal there are only two choices, 0 or 1, then the signal is most distinguishable from noise. If coding consisted of a different amplitude for every digit or every letter of the alphabet, then the slightest noise added to the amplitude might make the level for a "J" indistinguishable from that for a "K." Efficiently coded signals are very susceptible to noise. One strategy against noise is to use redundancy in the coding, perhaps by sending each message twice or three times.

A concept from physics called *entropy* is related to information. Entropy is a measure of disorder, so that decreasing entropy represents increasing order. Information is rather orderly since it represents choices among specified alternatives. The entropy of information in bits is the minimum number of bits needed to specify the information uniquely. If the information is a character of the alphabet, for example, binary coding must assign at least five bits

per letter (six for upper and lower case). Five binary digits represents $2^5 = 32$ alternatives. But there are many ways to code letters and words more efficiently, and the lowest entropy has been found empirically to be about one bit per character. Computers must assign eight or even 16 bits per character and even store that much. Sometimes it is possible to calculate theoretically the minimum entropy (highest efficiency) of coding a certain kind of information.

14.3 THE RELATION OF FREQUENCY AND TIME FOR SIGNALS

Physically, information consists of energy fluctuating in time. The more monotonous the variation in energy, the less the information carried. A minimum energy is needed to transmit and process information. If the information is digital, the minimum energy per bit is given by:

$$E = kT \ln 2$$

where *ln* is a base e logarithm.

To send information across space, energy must actually be delivered from here to there. To receive information one must wait, first for the speed of light to deliver it, and second for the sequence of bits all to arrive. The amplitude of any wave has been shown to be related to the intensity (or energy) that it can carry.

The relation of the frequencies of the waves needed and the time structure of the information is of special interest here. A carrier modulated by a complex signal is equivalent to transmitting many frequencies together. Information can be considered in two different *domains*, time and the frequency. Just as time and frequency are related reciprocally, the two domains are related in a complementary way. The time domain may also be called the real-time signal, and the frequency domain is its spectrum. Figure 14-5 shows several side-by-side comparisons of signals graphed in time and their frequency spectra. A smooth sinusoidal signal that never stops represents just one frequency, and so it appears as a thin line on the frequency graph. A sine wave that starts and stops like a pulse is represented by a limited but continuous set of frequencies graphed as shown. The frequency of the carrier does not stand out in the spectrum.

A regular series of pulses carries little information (like sending 11111111111 forever) but nevertheless has a complex frequency spectrum. Predominant is the frequency representing the repetition rate of the pulses. Also present is the frequency related to the narrow time width of the pulses. To obtain the fast rises and falls requires some very high frequencies. A single pulse contains a wide range of frequencies in continuous bands up to infinity. Some graphs of spectra have relatively smooth envelopes as shown in Fig. 14-5. The spacing of bumps in the envelope is constant and related inversely to widths in the signal.

Experiment 14-1

To hear the spectrum of a pulse in a practical way with sound, clap your hands in a large, highly reflective room. You should hear a series of musical tones corresponding to the cavity resonances of

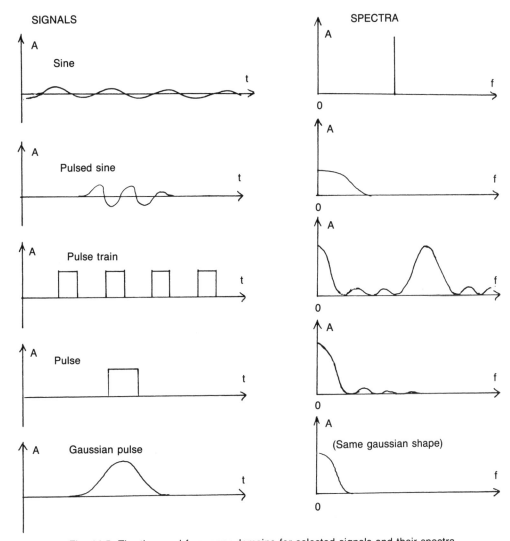

Fig. 14-5. The time and frequency domains for selected signals and their spectra.

the room. They are evoked because the pulse includes those frequencies and excites them. ■

The relation between a pulse and its spectrum should evoke a recollection from optics. If the pulse represents the pulse of light that passes through a hole, then the spectrum seems to be shaped like the diffraction pattern that results on a distant screen—the graph of intensity, that is. The relation is not accidental but very basic. More will be made of this later. To be more accurate, the graphs should show negative frequencies (simply a 180° phase shift) in order that the curves represent the symmetry of diffraction patterns.

The mathematical connection between the time and frequency domains is made with a function called the *Fourier transform*. Its explicit form need not be shown in order to continue using the concept. The process of finding what frequencies are needed to carry a certain pattern of information in time is called *Fourier analysis*. As one might expect, there is one signal whose graph in time has the same shape as the graph of frequencies of which it is composed. This is the gaussian or bell curve shown in Fig. 14-5. Does the train of equal pulses and the accompanying spectrum recall any other well-known optical device? And what function in time has a transform resembling the two-slit interference pattern as a frequency graph?

14.4 PICTORIAL INFORMATION

Information in a picture seems at first to be different from information previously defined but can be discussed in the same terms. A television signal is constructed by scanning along rows of an image of a scene. The brightness at successive points along a row is encoded and transmitted. This is done row after row until the whole picture is encoded and sent. Because of noise and limited bandwidth, there is a minimum size of picture element that can be handled. This is called a *pixel*. Pixels can be thought of as small round, square, or rectangular dots. If you have used a computer screen, the display is broken into tiny pixels. The number of pixels horizontally and vertically should be about the same, as the eye

is sensitive to pictorial information about the same in both dimensions.

Television in the United States breaks pictures into 525 lines or rows vertically, with each line having about 400 distinguishable pixels along each horizontal row. Surprisingly, the vertical resolution is worse than one line, so that about 300 pixels can be distinguished vertically. One picture then has about 120,000 pixels. Thirty complete pictures are sent each second so that motion can be shown. Thus the pixel rate is 3.6 million per second. To obtain the bit rate, the encoding of information about each pixel must be considered. If 32 levels of brightness are distinguished (a grayscale with 32 steps), a four-digit binary number must be used. If there is color, more choices are to be made for each pixel, and an eight-digit binary number might be needed for each pixel. The total bandwidth needed is about eight times the pixel count or about 30 million Hz. Present day television uses 4.5 MHz for color pictures, so clearly this analysis of information is overestimating the actual need. The difficulty is traceable to the inefficiency of digital encoding. While introducing noise in the picture, amplitude coding allows much more information to be packed into the same bandwidth. Any move to digital television would require a sophisticated coding scheme to fit the existing bandwidth, and several have been developed.

In comparison with television, the pixel content of modern color motion picture film (or 35 millimeter slide film) is upwards of ten million (the eye can distinguish ten million separate dots over the screen). The wide range of brightness and variety in color brings the total bit rate to nearly a billion per second when motion is included. It is astonishing that the eye can take in information at this rate. Actually, the eye scans the scene in still or motion pictures, examining details with the fovea a little at a time. Also there are usually broad patches of very nearly the same color and texture—large, redundant areas of pixels—that the eye largely ignores.

A different approach to pictorial information is more pertinent to optical problems and makes direct use of Fourier analysis. A flat picture consists of information coded in two spatial dimensions. If the original scene is three-dimensional, the use of lenses with good depth of field produces a two-dimensional real image on a screen. To start, consider the picture shown in Fig. 14-6A. This picture varies in brightness B in one dimension only and is understood to continue to infinity in all directions in this way. A graph of brightness over distance has a sinusoidal shape as shown. In the horizontal spatial frequency domain f_x, the spectrum would have one line as shown, corresponding to the spatial distance d between maxima. The units of cycles per meter are sometimes used for spatial frequency. Vertically there is no brightness variation. Depending on where one looks, the vertical brightness is constant at maximum, minimum, or in between as graphed in the inset. The spatial frequency is zero (analogous to direct current in electrical circuits).

Consider a picture composed of one vertical bar as in Fig. 14-6B. This is analogous to a spatial pulse, and its spatial frequency content in the x-direction is expected to be a broad band centered around zero as shown, with side lobes of lesser amounts. Vertically (y-direction) nothing varies. If the picture is a series of these vertical bars as in Fig. 14-6C (but extending to infinity), the brightness graph is a series of spatial pulses, and the spatial spectrum has a regular series of lines, starting with one at zero. The significance of the spatial frequency of zero is that it represents the average brightness of the picture (just as a series of electrical pulses has a net DC component). From far away, this picture would appear uniformly gray. Closer up the basic spatial frequency (call it f_0) corresponding to the bar spacing d is apparent. Fourier analysis applied to the edges of the bars would require some very high frequencies, and these appear as a regular sequence of lines up the spectrum (much like the orders from a diffraction grating). If the sequence of bars were not infinite, the lines would broaden to peaks and there would be small bumps between them. If the

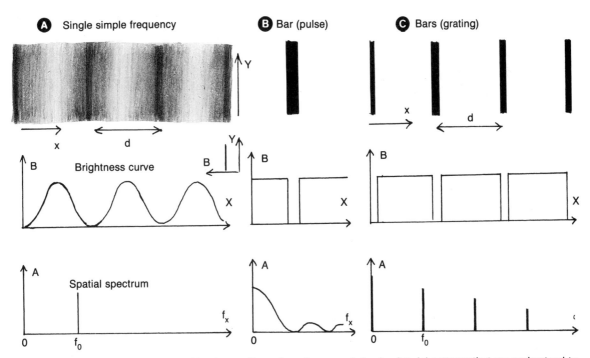

Fig. 14-6. Spatial frequency composition in two dimensions for several simple pictorial patterns that are understood to extend to infinity in all directions. Dark lines signify zero brightness, and inverted brightness curves have the same spatial spectra.

edges of the bars were blurred (poor contrast), the higher spatial frequencies are diminished or missing, as if they were not passed by an optical system.

As will be seen in the chapter on vision, the use of sharp-edged bars is basic to the way the eye detects features in a picture. We see not the details of the bars but the sharply contrasting edges. If one stares at one point on the pattern of bars for a while, one can begin to sense that maybe the eye is seeing the basic spatial frequency f_0, as everything seems to fluctuate about that spacing in one's vision.

Consider a more complex scene, say a grove of trees. If one squints or otherwise blurs the vision, the most apparent thing is a series of vertical dark bars, the tree trunks. If these are regularly spaced, there is one spatial frequency for them, which the eye notices. If the trunks are irregularly spaced, then the analysis is more complex. If the background is light sky, the sharp contrast between background and dark trunks is most apparent as a series of edges. The spatial frequency analysis can be done for any series of changes in brightness, although the results are perhaps too complex to be any more meaningful than the scene itself.

In the more complex scene, there are also variations in brightness along the vertical dimension. These can be analyzed into another set of spatial frequencies. A checkerboard is a good example of a object that would have about the same brightness graphs and spectra in both dimensions. Here the physical analysis of the content of pictures probably parts company with the way the eye intrinsically analyzes them, as the eye does not seem to consider separate spatial variations for the horizontal and vertical.

Similar to a checkerboard, but having larger holes, is the object formed by crossing two diffraction gratings or two Ronchi rulings (Fig. 14-7A). This object has spatial frequencies in two dimensions patterned like the grid of dots obtained from crossed gratings. Figure 14-7B represents a three-dimensional graph of the spatial frequencies, with amplitude shown vertically. The array of humps is like the knobs on a waffle-soled sneaker (before they wear off). They are shown here as fat because the crossed grating is assumed to have a small number of elements. For infinite gratings, the pattern would consist of sharp spikes. If the gratings had soft edges (the opaque areas becoming gradually transparent),

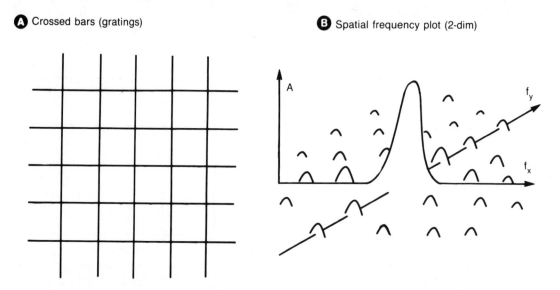

Ⓐ Crossed bars (gratings)

Ⓑ Spatial frequency plot (2-dim)

Fig. 14-7. Crossed Ronchi rulings or bars and a sketch of their spatial frequency graphs in two dimensions.

then the high frequencies would be suppressed and the pattern would have just a few peaks around the center.

14.5 OPTICAL PROCESSING

In the most general terms, an optical device is called an *optical transform* or simply transform. It is apparent from holography that a lens, for example, does not simply transform an object to an image but provides an intermediate stage. In simple situations where the object lies in one plane, the device causes an effect in another plane called the *transform plane*, and the image occurs in a third plane (Fig. 14-8). The transform plane, not considered previously, is where the lens forms the Fourier transform of the object. For planar light, the transform occurs in the focal plane. The results of an optical transform are identical to a spatial Fourier transform except phase information is omitted. With caution the two are not distinguished here. References (for example, Kallard) show many examples of Fourier or optical transforms, some quite beautiful.

To produce a transform of an object too large to show diffraction effects by itself, a lens is placed after the object with focal point at the object to convert object rays to parallel rays (Fig. 14-9). At the transform plane where parallel rays from all parts of the object interact, the Fourier transform occurs. A screen put there would show the diffraction pattern. Without the lens, the diffraction pattern will form at infinity. It is assumed that planar coherent light illuminates the object. To recover the real image, a second lens serves to transform the diffraction pattern and produce a focused real image on a screen. Two transforms in series are equivalent to no change. The second undoes the first. The two lenses must be high quality. The whole system is sometimes called a coherent optical computer.

Combining the concept of spatial frequency with this general approach to optical processing gives a fundamental result: a diffraction pattern is the same as spatial frequency analysis. In other words, diffraction gives the Fourier transform of an object. A lens can perform a Fourier transform either way. If one lens produces the transform of an object, a second lens can untransform it to make the image. The second lens must be large to capture enough of the higher spatial frequencies to form a good image.

An important application of these ideas occurs in *spatial filtering*. In the transform plane, all the information about the object has been converted to spatial frequencies in two dimensions. The graph of this information exists in midair and can be operated upon (see Fig. 14-9). Suppose, for example, that the object consists of crossed bars as before. This has high spatial frequencies, corresponding to the sharp edges. If a mask (the spatial filter) is placed on the axis of the pattern with a hole that admits only the low frequencies (the peaks near the axis, around zero frequency), then a sharply focused image of the crossed bars (obtained with a lens that transforms the transform back) will appear blurred. The edges are all soft. The brightness variations are gradual, almost sinusoidal. If only the central peak is passed, all information about the crossed bars is lost, and the only image is a uniform illumination. If the spatial filter consists of a narrow slit that passes a central row of the spatial frequency pattern, then the set of bars corresponding to that row will be formed as the image. Matching spatial filters can be made by putting opaque paper at the transform plane, observing the pattern, and marking places to be cut away.

The practical applications are many and growing. Suppose that some dirt is sprinkled on the crossed bars in the preceding example. If a spatial

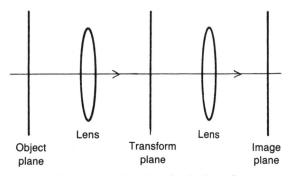

Fig. 14-8. General consideration of optical transforms performed by lenses.

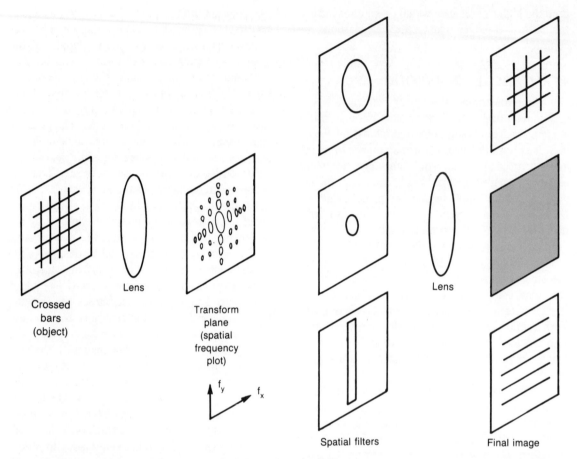

Crossed
bars
(object)

Lens

Transform
plane
(spatial
frequency
plot)

f_y

f_x

Lens

Spatial filters

Final image

Fig. 14-9. A lens transforms an object (crossed bars) to a diffraction pattern in the transform plane, where selected spatial frequencies can be filtered out with special apertures in the same plane. A second transform gives the resulting, altered image.

filter consisting of a regular grid of *opaque* spots painted on clear plastic is used, sized to match the transform, then all frequencies relevant to the bars are filtered out and the remaining frequencies corresponding to the dirt are passed. An image of the dirt is obtained after a second transform. This approach could have been reversed, removing the irregular frequencies of the dirt and keeping the regular pattern of the bars. The method is especially good for removing regular but unwanted parts or flaws from a picture, since the number of frequencies to be removed is small when the pattern is regular.

The methods of character recognition follow quickly from spatial filtering. The basic idea is that the spatial frequency pattern is the same regardless of the location, size, or small differences in the features of a particular class of objects. For example, the letter "R" in various typestyles and sizes has almost identical transforms. (The more the eye has to study a character to determine what it is, the less likely it will have the same transform.) All the letters "R" could be picked out of a page of text by using a lens to make the transform, then placing a filter for the frequencies of "R" there. The final result is an image showing each "R" in its place without any other letters of the text. Alternatively and most peculiarly, if a stencil of "R" is placed in the transform plane, a bright spot is obtained in the image for all occurrences of "R" in the text. Although this method sounds promising, many difficulties arise as more complex applications are studied. Greater hope may lie in the use of holography.

A major problem in physics and in the technology of materials is to decipher the structure of a material sample from the diffraction spots produced when x-rays are passed through the sample. While mathematical methods exist for calculating crystal structure from diffraction patterns, the amount of computer time has been forbidding. The x-ray diffraction pattern is in effect a hologram of the crystal, although phase information must be added separately. With computer assistance a Fourier transform is constructed from it which can then be viewed with laser light to see an image of the crystal structure.

A more ambitious pattern recognition system uses holograms that store many images of objects of various sizes, for example, different faces. When an image of an unknown face is introduced into the laser-based system, the best match with an existing image is automatically outputted. Variations in size are irrelevant, and the method can work with only a part of a facial image. The system may work similarly to the way humans rapidly recognize faces in what may be a hologramlike memory storage.

The deblurring of photographs can be done with no more information than the blurred photograph itself. What is needed is the effect of the blurring on a known point (a dot) in the photograph. A hologram of this blurred point then serves as a deblurring transform. The blurred photograph is then transformed to its Fourier transform by the lens method and combined with the deblurring transform. An improved image is sent to the image plane. The blurring need not be due to photography or a lens system. Any image viewed through any obscuring medium can be corrected if a hologram of a point object in the medium is obtained (Fig. 14-10). Coherent illumination and reference beam are needed as usual. The hologram is used as a compensator through which the original object is viewed.

An opposite approach to pictorial information processing is conventional use of a conventional computer. The input image, perhaps a photograph, is first digitized—that is, scanned with brightness measured and then coded as binary numbers in a computer file. Many kinds of operations on individual pixels, selected groups, and the whole picture become possible. For example, if a blurred photograph is received from a spacecraft, and the identical optical apparatus is available for a special test photo, the optical transform of the apparatus can be measured and digitized. Then the inverse transform can be calculated—the one that would deblur—and it is applied to the blurred photo.

The degree of poor focus of an optical system can be measured with special test patterns. Com-

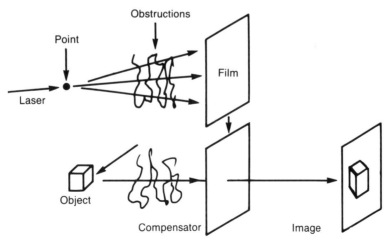

Fig. 14-10. A hologram of a point object can be used to correct the image of a complex object.

puter processing can also enhance contrast and remove unwanted regularities. Picture processing takes much time, hence the attractiveness of purely optical methods that operate on all pixels simultaneously. Computer calculated and plotted false color images are in wide use in nearly every branch of science and technology now. Picture processing blends with the vast field of computer graphics and includes attempts to mimic human comprehension of the content of pictures by processing components of pictures such as the outlines of objects.

In the converse computer process, pictures are generated in a series of steps. First the three-dimensional object is represented mathematically (or with data). Then a two-dimensional view of it is computed, perhaps in perspective. At this stage only outlines of the object are considered and all are seen (a wireframe view). Then lines not visible from a particular viewpoint are removed and visible surfaces are identified. The color, shading, and highlights of the surfaces may then be computed from specified lighting. Refinements include shadows, colored reflections from nearby objects, color changes in specular reflection, surface textures, and smoothing of lines. To go this far toward a realistic computer image takes much computing time and motion pictures with good realism cannot be generated in real time yet. The uses of this process are many for science, engineering, and training, as well as for entertainment.

15

Nonlinear Optics and Other Modern Applications

This chapter introduces you to nonlinear optics—phenomena and applications resulting when high-intensity light changes the refractive index and speed of light of a medium. The laser is often featured here, and holography is a necessary tool. Nonlinear effects permit an expansion of optical information processing capabilities. This chapter also surveys some of the applications of optics to modern problems in many fields of science and technology. All advances depend on new materials, and the general article "Optical Materials" by Glass is an excellent review of recent developments.

15.1 NONLINEAR OPTICS

In linear optics, two light beams can cross through each other with no changes in the beams, although interference might be detectable in the region of interaction. In nonlinear optics, the beams may permanently change each other, and new frequencies of light are produced along with the original ones. *Nonlinearity* occurs when the index of refraction is changed by the light.

The intensity of laser light allows an effect to be seen that is not accessible otherwise. The strong electric field is so high that when the light shines through certain crystals, the electrons respond nonlinearly to it. Typical electric field for an electron loosely bound to an atom is about 10^{10} volts per meter. An intensity of about 10^{17} W/m^2 is needed to achieve this, or about 100,000 watts on a square micrometer. Instead of vibrating sinusoidally, the electrons exceed the range over which they can mimic the light wave. They may oscillate more one way than another or may be in effect jerked from one position to another (Fig. 15-1). In more extreme cases, they are ripped from their atoms. When electrons respond less smoothly than normal, they emit light of several frequencies. In the first approximation they emit light of two and three times the frequency of the incident light. If strong infrared light at $3(10)^{14}$ Hz is shone into a crystal such as ruby, light at $6(10)^{14}$ Hz (blue) might emerge. The double frequency light is second harmonic light and has practical uses.

A little must be said about optical parametric amplification. When a power (pump) laser beam and a signal beam enter a nonlinear material, some of the power is transferred to the signal beam. A third beam (the idler) is also generated which has a frequency that is the difference between the pump and signal frequencies. This is also called three-wave mixing, because three light beams are involved (Fig. 15-2A). A nonlinear amplifier mixes the input signals together in the sense of multiplying them. A result in trigonometry for any sinusoidal curves is that multiplying two waves together results in waves with frequencies that are the sum and difference of the original frequencies. The multiplication occurs because the response of the nonlinear crystal is partly to the square of the total electric field. Only the difference frequency occurs when the parametric amplifier is tuned to resonate at the difference frequency. Because of the resonance, energy is transferred from the pump beam to the signal and idler beams. This constitutes amplification. The amplifier is tunable because a parameter such as pump intensity can change the index in the cavity crystal to cause resonance at a variety of frequencies.

Mathematically, the nonlinear multiplication of waves appears thus for intensities:

$$(E_1 \sin \omega_1 t + E_2 \sin \omega_2 t)^2 = E\$_2^2 \sin^2 \omega_1 t + E_2^2 \sin^2 \omega_2 t + 2E_1 E_2 \sin \omega_1 t \sin \omega_2 t$$

where ω (omega) is an angular frequency.

If two intense laser pump beams of different (or same) frequencies are shone in opposite directions into a nonlinear crystal such as lithium niobate (Fig. 15-2B), along with a signal beam, two different beams emerge with new frequencies that are the sum and difference of the signal and pump frequencies. The arrangement is called a four-wave mixer. Three frequencies are mixed together, although the same frequency is often used for all three beams. Energy is transferred to the signal beam to form an output beam that emerges parallel with the signal beam.

When the output and idler beams are combined, squeezed light results, with some of the natural noise squeezed out. (Mathematically the two beams are a *sin* plus *cos* quadrature representation of a sine wave with phase, and one component is amplified.) If the signal is nothing (natural noise), then light with less noise than statistically allowable emerges at the expense of more fluctuation in phase. Nonlinear be-

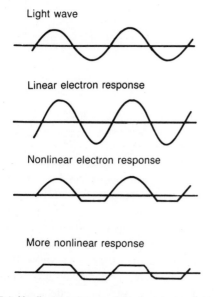

Light wave

Linear electron response

Nonlinear electron response

More nonlinear response

Fig. 15-1. Nonlinear response of an electron to a light wave.

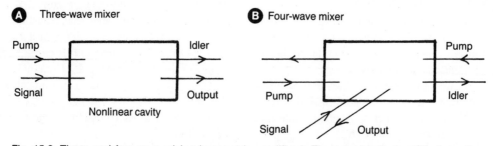

Fig. 15-2. Three- and four-wave mixing (parametric amplifiers). The output is the amplified signal.

havior of materials has more general effects than just the mixing of waves. Any property of a material changed by a wave or field in the material constitutes a nonlinear response. The material in turn changes the wave or field.

Another very sophisticated technique that requires a material to be used nonlinearly is the formation of a transient or virtual hologram—this is, a temporary interference pattern stimulated in a material by interacting laser beams and present only while the light is present. A transient hologram in the form of a grating (virtual fringes produced from the interference of two plane waves) has been found useful for analyzing very short light pulses. the method is another version of four-wave mixing.

15.2 OPTICAL PHASE CONJUGATION

A problem that can be solved in principle is the automatic restoration of coherence in a laser beam that has passed through a fluctuating material or a material that is not homogeneous, such as the air or translucent plastic (Fig. 15-3). Or a second laser cavity can be activated. Wherever a coherent beam can be made to pass, so can a clear image be transmitted. The *phase conjugation* in this process is to reverse the path of the light used and therefore its phase. Reversed light automatically retraces its path back to coherence. The method cannot undo the effect of diffraction, however.

The key element is a piece of nonlinear material used as a phase conjugator. It can be a crystal, glass, liquid, or gas. High intensity and a sensitive material are required, so that the scattering of light

in it changes the material and causes more scattering. Wherever the intensity is highest, the scattering is strongest. Interference between incoming and outgoing light causes more in-phase scattering. This scattering is called stimulated Brillouin scattering and begins at an intensity of about a million watts per square meter in most materials. The incoming beam sees a reflective medium that exactly matches its wave front and thus is almost literally turned around. A mirror cannot be used to obtain the reverse beam because a mirror only reverses a plane wave incident headon, not a wave front varying from passage through a complex material. A phase conjugate mirror sends a light beam back in the direction from whence it came, like a retroreflector. Simple retroreflectors used on highway posts and signs are low-resolution phase conjugators whose performance is limited because of the large size of the individual elements. Optical phase conjugation resembles active or real-time holography because an instantaneous and ever-changing hologram is formed in the nonlinear scattering medium.

Phase conjugation was first achieved with Brillouin scattering in 1972 at Lebedev Physical Institute in Moscow. It can also be done with a four-wave mixer as described later. The phase conjugator is somewhat slow in its response and is able to cancel fluctuations in the laser beam due to the laser itself. Besides improving laser beams and compensating for atmospheric fluctuations, optical phase conjugation can help track moving objects, encode messages, and compare images. It is an automatically adapting system. One tracking application is to aim

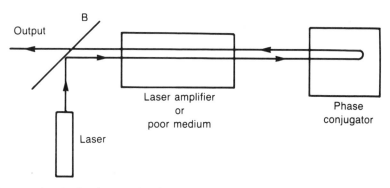

Fig. 15-3. A phase conjugator (conjugate mirror) used with a laser cavity or medium that distorts a wave front.

a high power laser at a moving fusion pellet which is first detected with a low power laser. It can image an object without lenses, as in placing a photolithographic mask on a silicon chip for microcircuit etching. The tracking and compensating abilities of a phase conjugator were perhaps best illustrated when an ordinary kitchen spatula was substituted for the usual precision mirror at one end of a laser made with a phase conjugating crystal. Regardless of small motions, an intense laser beam occurred using the spatula (see reference by Pepper).

A phase conjugator assists holographic memory in matching an input image to its stored images. A phase conjugator can restore light pulses to their original form. During passage through an optical fiber, the various frequencies in a pulse spread due to their different speeds, with the shortest wavelengths arriving last. The phase conjugator can exactly reverse the order of the frequencies in the composition of the pulse. When the reversed pulse is added to the original, the result is a narrow pulse. Sequences of pulses can be reversed, too (Fig. 15-4). When reflected in an ordinary mirror, the first pulse in is the first out. A phase conjugator exactly reverses the whole sequence, preserving the original spatial order of the pulses and reversing their order in time. For this application, the conjugator must be deep enough to hold the sequence of pulses at one time.

15.3 PHOTONICS OR GUIDED-WAVE OPTICAL CIRCUITS

A different application of optics to calculation is the hope of an optical computer that uses light guides instead of wires and that processes optical information instead of electrical. Electric current would be replaced with light, preferably coherent laser light from laser diodes. The construction of integrated optical circuits is called *photonics* or *guided-wave optics*. Many optical devices are analogous to electrical ones. The most needed device is an optical transistor or switching element. This is not an electrical device that is controlled by light—although those are needed, too—but a purely optical device that controls a light output by means of a light input. The heart of any computer has active devices that can combine two or more signals (usually pulses) to form an output pulse that is logically related to the input pulses. Computers, including planned optical ones, do most information processing in serial form. This is in contrast to the coherent optical computer (of the last chapter) that processes every pixel simultaneously in parallel without circuitry.

Progress has been slow in more than 20 years concentration on the photonic computer. The key element, a bistable optical device that can be switched to one of two states, requires careful use of nonlinear materials and accurate high intensities of light. Amplification with delay is needed to provide stability for the two states and to restore weak light levels to full strength. *Fanout* is another important requirement. The output of a computing device must have enough power or intensity to share among many other devices.

A chief contender for the basic photonic transistor is a specially prepared Fabry-Perot interferometer (Fig. 15-5). Between its two parallel mirrors is a very nonlinear optical material. A power light beam with accurately set intensity is shone at

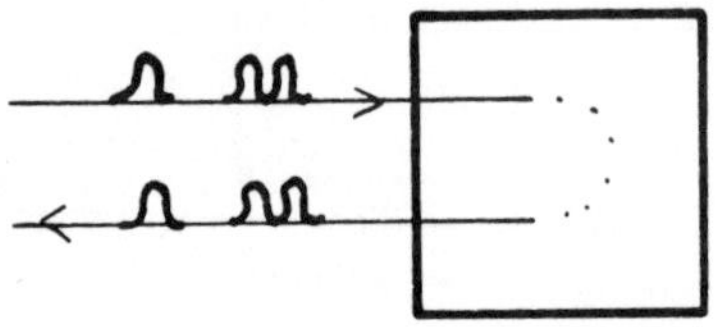

Fig. 15-4. A conjugate mirror reverses a pulse sequence in time but not in space so that the first pulse in is the last out.

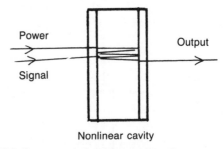

Fig. 15-5. A proposed photonic transistor that controls one laser beam with another uses a nonlinear medium in a cavity.

the proper angle into one silvered mirror. Most of it is reflected, but some enters the cavity where it reflects back and forth. The spacing and material are chosen so that reflections of the wave are out of phase, and a standing wave is discouraged. If a signal beam is also shone into the device, then the total intensity crosses a threshold level and the material goes nonlinear. The speed of light slows down sufficiently that the cavity is resonant, a strong standing wave builds up, and light emerges from the other side. The cavity stays locked in this mode until the signal beam is removed. The transmission state is stable since it is immune to small signal variations. When the signal is gone, the cavity goes back to its original stable state of not passing the power beam.

An important small delay in the switching between states is provided by the tendency of high intensity to hold the device in resonance. The device also amplifies, because the output can be many times the signal strength. It pumps power from the power beam to the output beam when commanded to by the input signal. It can operate about a trillion times per second, over a hundred times faster than electronics. Unlike an electronic device that will lock into either state even if the signal is removed, this device must relax when the signal is removed. The only other way to bring it down from the transmission state would be to reduce the power beam momentarily. Clearly the angles and intensities of the incident beam must be set precisely. What simple devices will adjust them has not been settled yet. This device must be made very small and in vast quantities. Ways to make the basic logic circuits AND, OR, and NOT have been worked out. Any sort of computer can be assembled from these. The mirrors are simple silvering on parallel faces of a material. The light is infrared, to which a material like indium antimonide is transparent. This photonic transistor does have the possibility of multiple stable states, unlike simple electronic devices. However, the high intensity light needed presents problems of overheating and thermal stability.

Another type of photonic switch consists of two light guides (essentially, optic fibers) formed in lithium niobate crystals and separated by a small gap

(Fig. 15-6). When an electric field is applied between the crystals, the index of refraction is changed and the light can leak from one to the other. It then takes a different output path. This switch operates as fast as electronics can drive it (not quite 10 GHz). It is also better for pulsing a laser beam than direct electrical switching of a laser diode because pulses are cleaner and shorter. The switch is not bistable and requires continuing electronic control. In principle, such switches can be assembled to form a computer. This particular device has proven easy to fabricate, has very low losses, and needs little voltage to control as compared with similar optical devices made with semiconductors. A major challenge for photonics is to build large circuits with very tiny elements in an integrated fashion—that is, all together on a single substrate of material, perhaps including electronic devices. Problems include thermal stability, optical damage, coupling, and inadequate nonlinearity (too small an effect from a large stimulus).

Another approach recently proposed for a light-switch is the use of a laser to change the energy levels of gallium arsenide to absorb a shorter wavelength. When the absorption level is shifted, then another laser beam will either be absorbed or not, depending on its wavelength. This switch will work fast (less than a picosecond). High intensity is required, however. A related approach involves multiple quantum wells in a stack of very thin layers of semiconductor. Again there are heating problems as each laser pulse changes the index of the medium by depositing heat in it.

One type of integrated optical circuit explored has optical components constructed in thin films on a substrate. The possible components include the equivalents of prisms, mirrors, lenses, active

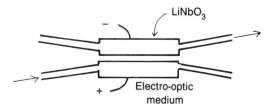

Fig. 15-6. A guided-wave switch uses the electro-optic effect on lithium niobate to control the leaking of waves across a gap.

switches using the Kerr and Faraday effects, filters, and polarizers. The optical "wiring" consists of light guides in which the light reflects between top and bottom surfaces of a thin transparent film many times as it moves forward. The film index is chosen to promote good internal reflection. A region where speed is lower and refraction can occur is a region with thicker film so that the light takes longer to travel. The film must be very smooth or the light is quickly lost scattering from defects. Different light waves can be kept separate in the same film by introducing the light at different angles. A thin film infrared laser is made from two mirrors, each of which consists of rulings on the surface spaced a half wavelength apart. Constructive interference provides the equivalent of reflection. A polarizer is made by depositing a conductor on the surface. Pulse modulation can be accomplished with parallel light guides controlled by an electric field as before. Practical implementation outside the laboratory has not been achieved yet.

15.4 DATA TRANSMISSION BY OPTIC FIBER

Before lasers were known, the use of optic fibers to transmit information was explored. As discussed in Chapter 6, optic fibers are thin with double walls to promote total internal reflection. Modern research efforts have gone into making extremely low-loss glasses and in using a nonlinearly distributed index of refraction to control light pulse shapes. Fibers should be round or light waves can be trapped in narrow corners. The lower limit to fiber size is near the wavelength of the light used when the fiber acts as a waveguide with different modes of transmission. The light then emerges in patterns much like those in a multimodal laser.

The first modern laser-fiber communication system was a 1.5-mile, 24-fiber telephone and video link

in 1977 in Chicago. Each fiber carried 45 million bits per second (pulse modulated). A laser is not essential for this system. An LED (light emitting diode, similar to a laser diode but used below the threshold for coherent radiation) is sufficient. LEDs have proven very durable and laser diodes are expected to be. The active junction in either is very small, hardly larger than the fiber. The LED light spreads in all directions and a lens is not suitable for collecting it. For best coupling, the fiber can be inserted in a hole in the active LED material. Much more laser diode light can be directed into a fiber with a tiny lens. The bandwidth of laser diodes is about 5% that of LEDs. Different frequencies travel at different speeds in a fiber, so a pulse tends to spread out. Laser diodes permit much more narrow pulses to be used. Both sources operate in the ideal 0.8 micrometer range. Recent developments are indium phosphide and indium gallium arsenide phosphide (InGaAsP) lasers and detectors in the 1.3 to 1.5 micrometer range.

Pulse modulation is not absolutely necessary—there are ways to modulate the amplitude of light continuously—but it does bypass many problems of distortion, noise, and loss. After some kilometers of fiber, the signal is so weak and dispersed that it must be amplified at that point (Fig. 15-7). Most signal restoration amplifiers are electronic, so the weak signal is converted temporarily to electronic by a detector, amplified, and then reconverted to narrow light pulses. A way to control dispersion is to use graded-index fibers (Fig. 15-8). The index decreases with distance from the center of the fiber, and it decreases more rapidly the farther away the light wanders. Light travels faster near the outer parts of the fiber. The result is that light waves are reflected gradually and speeded up rather than reflected abruptly, losing time. The travel distances of waves at various shallow angles are more nearly the same. This technique reduces dispersion up to

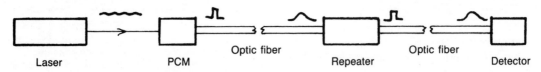

Fig. 15-7. An optic fiber communication system.

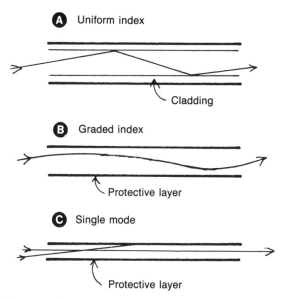

A Uniform index

Cladding

B Graded index

Protective layer

C Single mode

Protective layer

Fig. 15-8. Three kinds of optic fibers. The single mode fiber is achieving dominance in applications.

a factor of 100 and increases the usable frequency of a fiber in the same proportion. Single-mode fibers (Fig. 15-8C), which do not rely on reflection but carry only axial rays, are emerging as the preferred approach to push pulse rates to ten billion per second over 100 kilometers.

Two kinds of light detectors are a photovoltaic PIN (for p-type, intrinsic silicon, n-type) detector whose name implies a three-layer device, and the avalanche photodiode. Both of these semiconductor detectors introduce noise in proportion to the bandwidth attempted, but the avalanche detector is quieter. Operation is limited in detectors that convert light to electrical signals to under 100 million bits per second, although fibers can be designed for higher speed. Progress is being made at two billion bits per second with indium gallium arsenide avalanche detectors. Presently, repeater amplifiers are electronic, but thin film optical amplifiers may become possible. Despite pulse techniques, fiber systems do have significant bit error rates. The rate may become as low as one in a billion bits, barely tolerable for computer data.

The most crucial advances are in fiber clarity. Reducing fiber losses is more effective than increas-ing light transmitter power. Damage-proof cladding is necessary to prevent losses, too. Fused silica glass is a choice material since it has losses not far above the Rayleigh scattering limit throughout the visible band and into the infrared. The theoretical minimum loss in almost any fiber material is strongly dependent on wavelength, and the minimums of less than 0.1% loss per kilometer are falling between 1.5 and 6 micrometers of wavelength for materials studied thus far. Impurities such as water must be reduced to extremely low levels. Present achievemnets are below 1% loss of light power per kilometer. In 100 kilometers, almost 40% of the light remains (from calculating 0.99^{100}. A practical problem is that light cables are not manufactured full length and must be spliced. This is a difficult problem when each fiber is much finer than hair and cannot be jumbled. Moreover, losses are introduced at connections.

A major advantage of optic fiber communication is the lack of signal leakage from one channel to another and the impossibility of electromagnetic interference with the signals. Economically, optic fibers are expected to become cheaper than wires, but even expensive fibers have an advantage because of the hundredfold greater channel capacity. The major fiber project underway at this time is TAT-8 (Transatlantic Lightwave System). It is 6650 kilometers long with 130 optoelectronic repeaters under the ocean and is expected to function about 25 years. Practical repeater spacing is about one every 50 kilometers. The bit rate is almost 300 million per second. Similar systems are being installed on land.

Short fiber runs such as a meter are proving useful in medical applications and almost anywhere data must be transmitted within or between equipment. Information that originates optically, such as the glow of a headlamp, is especially likely to be carried by fiber. Remote chemical sensing is another important application of optic fibers. Industrial processes involving radioactivity, toxic chemicals, high temperature, and other hazards can be monitored by sending the spectral information about the materials involved down optic fiber to be analyzed remotely. Temperature itself is monitored rapidly using rare earth phosphors in an *optrode*. The information originates as fluorescent emission from molecules, perhaps stimulated by ultraviolet sent down the fiber. Pollutants can be monitored in waste dumps and aquifers by this method. The location of heat or certain chemicals can be ascertained anywhere along the fiber if both ends of the fiber are accessible and monitored. Optrodes are now being inserted in blood vessels to monitor the blood chemistry of patients, too.

15.5 OPTICAL INFORMATION STORAGE

Optical storage can be holographic, as previously described, or digital. Because of the small size of light wavelengths, extremely dense storage is possible. Holograms were one way to achieve the highest density, using all parts of the film. A disk-type storage stores binary data as dark and light, or reflective and transmissive, spots. Data is entered by burning holes in the disk or burning off a reflective coating with medium laser power (Fig. 15-9A). Data is read out point by point by deflecting a focused laser beam to each of many tracks and measuring the amount of reflected or transmitted light. The readout method is opto-mechanical. A rotating turntable moves the disk, while a swiveling mirror scans the focused laser beam across the tracks. A laser of only 20 mW power can burn data into a track moving at 10 meters per second. A semiconductor laser can provide pulsed power at this level. Only about 10^{-11} second is needed to make each hole, and the holes are about a micrometer apart. About 10^{11} bits can be put on one disk. Data reading rates can be sped up to hundreds of millions of bits per second by reading many tracks in parallel. "Jukeboxes" containing hundreds of disks and a player have been developed.

Laser etching produces a permanent disk which cannot be altered, a virtue in some applications. When there are frequent changes in stored data, the capacity of the disk is so large that a new updated file is created and the old one saved, too. (Sometimes the original data is needed, too.) Work continues on ways to make an erasable disk that can be written and rewritten indefinitely. One approach is to use a magneto-optic film (perhaps gadolinium cobalt) on the disk. Data is stored by orienting small magnetic domains in the film up or down (Fig. 15-9B). When light passes through a domain, its polarization is rotated one way or the other. Data is detected via the polarization of the reflected light. When a magnetic field is applied, heat from a laser pulse can reverse the orientation of a domain. This sounds like magnetic recording, but there is no need to focus magnetic fields in tiny regions, recording and reading out are much faster, there is no wear on the disk, and the data density is much higher. Conventional magnetic disks have a billion bits of storage in the best systems, but an optical system can exceed this by a factor of a hundred.

Optical video storage was on the market for several years. The method resembles nonerasable digital recording except that data is encoded in semi-analog form rather than binary, using holes or pits of differing length along disk tracks. It is suscepti-

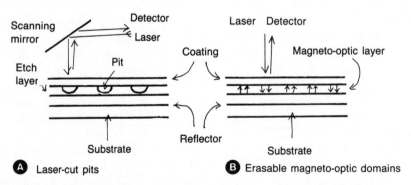

Fig. 15-9. Optical recording and reading by laser on a disk with laser-etched pits (nonchangeable) and with magneto-optic domains (changeable with heat and magnetic field).

ble to dust. Videodisks were developed for mass production, with a master disk being made directly from the laser-etched disk. One disk can hold one hour of video signal (both sides), equivalent to about 400 billion pixels. The error rate is about one in 10^5, not tolerable for computer data, even if error correction procedures are used. Adaption of the method to true digital recording as described above is expected to be technically and commercially successful.

15.6 SCHLIEREN PHOTOGRAPHY

Light passing through mixed or heated media undergoes variable refraction as it passes through regions of different density. The effect is called *schlieren* interference (from German for inhomogeneity). The effect can be orderly, as in a mirage, or disorderly as in the twinkling of stars. The phenomenon may be very rapid, such as the air currents at the tip of fan blades. High speed photography is usually—but not necessarily—used with schlieren optics. If the index of refraction varies in one direction, as in the atmosphere, then light will follow a parabolic path through the medium.

Simple schlieren image-forming systems use knife edges. The method was developed in 1864 by A. Toepler to test lenses. The method is in effect spatial filtering of interference patterns. Regions of different refraction are transformed into contrasts of light and dark as the light is either blocked by the knife edge or not. The objective of schlieren photography is to form an image of the variations in index and therefore the inhomogeneities of the medium itself. This can be done with a microscope or on a larger scale. The optical components must be very good, or their imperfections will appear. Color systems may provide more useful—certainly more vivid—information. In black-and-white or color, schlieren is often used to examine high-speed phenomena in air. Very intense white light sources are needed. They can be pulsed but must be brighter and faster than a xenon strobe light.

Figure 15-10 shows a common modern arrangement using reasonable-cost mirrors instead of extremely expensive corrected lenses. Bright white light (for color) or nearly monochromatic light is

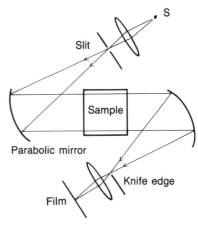

Fig. 15-10. Schlieren photography of density variations in a transparent medium.

used. Parallel rays from a slit source illuminate the sample chamber. The background light is brought to a focus at a knife edge oriented parallel with the slit. If variations in the chamber refract the light, an image of the change is brought to a focus that bypasses the knife edge and is focused on film. There are techniques for using color to code the density variations for making measurements.

15.7 ENERGY FROM LIGHT

The energy in light such as sunlight can be extracted by several methods. Mentioned here are only those most pertinent to optics. The energy consists of about 50% infrared, 45% visible, and 5% ultraviolet. Since most clear materials transmit ultraviolet poorly, attention is focused on the infrared and visible components. The typical maximum available energy at the Earth's surface is about 1000 watts per square meter.

Trapping by the greenhouse effect is the simplest method. Near infrared and visible pass readily through most clear plastics and glass. When converted to heat radiation on surfaces, this far infrared cannot easily escape through glass although more readily through some plastics. An enclosed area thus becomes warmer and is called a solar collector. The unavoidable losses are large, or else temperatures approaching the surface of the sun are possible. Such temperatures are achieved when mirrors are used to collect the energy over large areas and focus it in a small region. For capturing lower grade heat, various parabolic reflectors have been designed which either must track the sun or provide adequate concentration without tracking. The latter are cylindrical parabolic, being curved in one dimension only and compatible with tube-shaped collectors.

A photovoltaic device is a semiconductor designed to capture visible light with electrons at a *pn* junction and convert it to electrical energy. The theoretical efficiency seems to be 29%, although about 10% is now achieved in practice. Materials are being improved, and efficiency is boosted by concentrating lenses which double or triple the light intensity on the expensive photovoltaic cell. Much more ambitious would be direct rectification of sunlight. Since light consists of randomly varying electric fields, tiny antennae and rectifiers would be needed to convert the electric energy to electric currents. On a larger but no more practical scale, polarized sunlight falling on certain nonlinear crystals produces a separation of charge which can be utilized as electric current. Because of the weak nonlinear effect, the efficiency is very low.

Another optical effect mainly intended for solar use is the *photochromic* effect whereby exposure to sunlight (mainly ultraviolet) darkens a glass with certain constituents. It can be used to shield buildings from solar input and provide passive cooling.

Photosynthesis is a very inefficient biochemical process in plants for converting light energy to chemical energy. Blue and red light are needed by the pigmented active molecule chlorophyll, and it absorbs strongly near 440 and 680 nanometers. Located in layers in chloroplasts, chlorophyll appears green because it reflects green and yellow light. Because molecules that interact with visible light are unusual, special molecular structures are needed to absorb red and blue. Other pigments in plants, algae, and bacteria absorb other parts of the spectrum to gather energy and have characteristic colors such as orange, blue, or red. The photosynthetic mechanism is very complex, and understanding has been slow. Two photons of light are absorbed, to release an electron from each of two kinds of chlorophyll. Several hundred chlorophyll molecules in a special trapping center are needed to capture one photon.

The net effect of a cycle of reactions is that carbon dioxide and water are converted to carbohydrate and oxygen. The oxygen comes from the water, so in effect water is decomposed. The carbohydrate represents stored energy. Nature has not been able to find more efficient ways to accomplish this. Seas of leaves are spread under the vast flow of photons from the sun, but so few are captured each second.

Experimentation continues in many laboratories to improve upon natural photosynthesis or to find simpler chemical reactions. Of particular interest is the separation of oxygen and hydrogen from water with catalysts and light energy. Photochemistry also concerns other useful chemical reactions that occur in sunlight.

Plants and animals warm themselves in the sun with several more sophisticated methods besides having dark colors at the appropriate time of the year. Butterflies orient their wings at the appropriate angle to reflect sun to their bodies with the iridescent thin film reflectors on the wings. Some arctic flowers have spherical or parabolic shape to focus sunlight. Many plants move to follow the sun simply to increase (or decrease) the solar intensity falling on leaves.

15.8 OTHER BIO-OPTICS

Natural creatures use light in many ways besides for energy and for vision (covered in the next chapter). *Bioluminescence* is the emission of light by organisms as diverse as bacteria, fungi, clams, worms, fish, and insects. The actual process is a version of chemiluminescence, where the chemicals are made and brought together by living organisms. Most dramatic are the fireflies of southeast Asia that gather in trees and blink greenish yellow in synchronization, the rows of green side lights and red headlights on the railroad worm of Brazil, the bluish glow of dinoflagellates (micro-organisms) in certain tropical ocean bays, the flashlight fish, and the luminous ink emitted by a squid.

In the deep ocean where no sun penetrates, it is not surprising that most organisms use biological light. The flashlight fish has light-emitting bacteria in a special organ beneath each eye and can turn the blue-green light on and off with a shutter. It uses the light to see, to communicate, and to confuse predators. Biochemical production of light is poorly understood except for the firefly. The firefly light organ goes through many chemical steps between the nerve signal to light up and the actual emission of light. There are just a few spectral lines. The gene for firefly light has been located and even transferred to certain plants, which will glow if supplied the active chemical luciferin. Firefly light can be called cold light because all of the supplied energy goes into the emitted light and none into heat. This is an efficiency that technology has yet to achieve.

Besides carrying on photosynthesis, plants have several other responses to light. Plants can tell when to grow, flower, and die according to light levels and light periods detected by the pigment phytochrome. The general mechanism is to tell when the periods of darkness (night) are long or short so that the new season may be assumed to have begun. Phytochrome is sensitive to red at 660 and far-red at 730 nanometers and distinguishes (somehow) between the two to determine the day and season. Germination is started by 660 nanometers red and flowering by 730 nanometers far-red. Each species has its own requirements, and lighting can be manipulated to cause germination, growth, flowering, and fruiting when needed. Fruits can attain their ripe colors only if sunlight is present to start the reaction to form the appropriate pigment. Plants and animals show many bright colors, with some of the most impressive being the iridescent ones due to interference films.

Phototaxis is the tendency of plants to grow toward light and of insects, motile bacteria, and other one-celled organisms to move toward or away from light depending on their preference. Plant roothairs grow away from light in order to seek the depths of the soil. Many plant stems, cells, and roots function as optic fibers to carry the simplest of light signals. The light detectors are usually spots of pigments related to those used in photosynthesis. The simplest creatures lack vision, but their light detectors may be thought of as rudimentary eyes which respond only to average light level and possibly direction of brightest light.

Snakes (pit vipers and some other species) have infrared sensors that are the two (or more) pits near the front of the head. These are most sensitive to the far infrared radiated by warm animals and other warm parts of the environment. Sensitivity is achieved by having the sensing membrane very thin. Because there are two or more pits, these snakes can "see" the environment stereoscopically and locate warm sources. A special visual mechanism is used by these snakes to compare the infrared and visual images to extract more information.

Some effects of light on the human body, and probably on many animals, are tanning, vitamin production, biorhythm synchronization, destruction of tissues and genes, and ability to see. Strong light can also prevent loss of calcium from bones. It might be assumed that light does not penetrate beyond our skin, but it penetrates some tissues several centimeters. Therapy with light has been found effective for several diseases.

The most obvious effects are the formation of vitamin D from compounds in the skin, and the injury to skin from overdoses of ultraviolet light that cause sunburn (erythema). Skin also grows thicker when exposed to sunlight. Tanning is the production of the pigment melanin and is caused by ultraviolet between 290 and 380 nanometers at sea level. At higher altitudes, more energetic ultraviolet reaches the ground from the sun.

Daylight, detected by the eye, causes a separate signal to be sent to the pineal gland in the brain, which controls production of the hormone melatonin (produced during darkness). The sensors seem to be separate from those involved in vision. This hormone then synchronizes the internal 24-hour biological clock of the animal with the diurnal cycle and tells when spring or other breeding season has arrived. In animals melatonin causes sleep, prevents ovulation, and affects other physiological cycles. Green light seems to have the strongest effect. The daily rhythm driven by sunlight has been established in humans, but no seasonal effects have been clearly demonstrated yet. Recently, exposure to morning sunlight has been found to prevent winter depression.

16

Visual Perception

The subject matter here changes from the technological to the biological. The end result of many optical applications is to provide new or better images to your eyes. How your eye physically produces images was discussed in Chapter 5. Now the interest is in the perception and interpretation of the images. There are many steps from physical image to mental image, many of them still unknown. But researchers have uncovered a vast number of clues using many different methods. Only some of the most basic results can be summarized here. The focus is on human eyes, but similarities are shared with many lower animals where research is more easily done. Different kinds of eyes have evolved in nature also. The concept of spatial frequency is reintroduced so that you need not have mastered it in an earlier chapter. Details of color vision are left to the next chapter.

An important point to keep in mind is that everything you see is ''appearance.'' We cannot perceive anything except that which your brain is physically prepared to have you see. Physical aspects of an image such as luminance (intensity) and color have psy-

chological counterparts not easily measured. Human seeing is a process much deeper than the seeing done by the most sophisticated instruments that record and process images in an objective way. Your vision is very subjective.

In the study of vision a variety of terms are used for the same visual concepts and components. Not all variations can be given here. For example, vision cells are also called photoreceptors, receptors, or visual sensors. To give home experiments would be to greatly expand the chapters on vision. Many excellent experiments are described or indicated in the references on visual perception and illusions, and some are possible with the diagrams given in this chapter.

16.1 RETINAL STRUCTURE

After a reasonably well-focused image is cast on the retina, image processing begins. At this point resemblance between eye and camera ceases. For the moment, assume that the image is motionless on the retina, and that only one eye need be considered. The retina is granular and composed of millions of living photodetectors in the form of cells. Prominent are the rod cells for low intensity and peripheral detection and cone cells for higher intensity detection of color and details. A few of each are shown in Fig. 16-1. The diameter of each is about a micrometer or two wavelengths of light. The slightly smaller cone cells number about 10 million and the rod cells about a hundred million. Despite their use for detail, the cone cells are much outnumbered by the rod cells to obtain more sensitivity. The rod cells are mostly near the edges of the retina. Toward the middle of the retina they are arranged in a hexagonal pattern with a cone at the center of each hexagon; this is the macula, about 3mm diameter.

Within the macula, the cone cells are packed densely in the fovea, with several thousand in an elliptical region about .3 mm across. The density is almost 100,000 per square millimeter there. These cones are smaller than usual. For sensitivity to low light levels rather than fine resolution, one must look

slightly off center so that the image of interest does not fall on the fovea. A surprising fact is that photons are supposed to shine all the way through several layers of nerve cells before passing through the intricate structures of the vision cells to be chemically detected. The retinal structure is therefore essentially transparent. The "wiring" or nerve connections to the vision cells are the first layer through which the light in the image passes on its way to the deepest parts of the retina. Behind the vision cells is a dark absorbing layer so that there is no scattering of light in the eye.

The optically active molecules consist of a dye combined with a protein, discovered by George Wald in the 1930s. The effect of photons on rods containing molecules of *rhodopsin* (Greek for rose-vision, also called visual purple) is to convert them rapidly to molecules of retinene. The low-energy photon causes the active part of the molecule to rotate to a different configuration. The rate of conversion is in proportion to the intensity of light. The store of rhodopsin in the cell is soon depleted, especially in

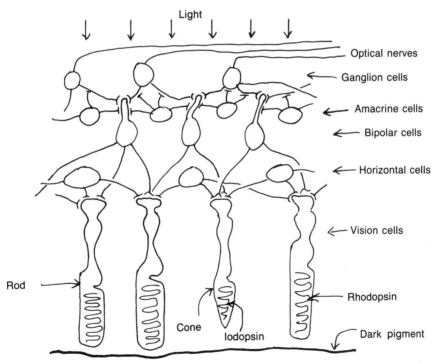

Fig. 16-1. Schematic of a small section of the retina showing six types of cells involved in detection and processing of visual information. There is no strict regularity in shapes or connections.

bright light, and low-level vision is lost until the supply is regenerated. This requires as long as half an hour and explains why vision is almost lost when one passes from sunny outdoors to a dark room. All the time one is in bright light, the sensitivity of the rods is suppressed by lack of rhodopsin (as well as overstimulation of associated nerve cells). Rod cells are also the first to be permanently damaged by exposure to excessively strong light, particularly ultraviolet. No more rhodopsin is formed, so night vision is lost.

The cone cells contain three kinds of photon-sensitive molecule called *iodopsin*, related to rhodopsin. Again, the optically active component is combined with a protein and folded into layers. Rhodopsin and iodopsin are pigments with their own intrinsic color, a reddish pink. Detailed knowledge about how rods work is partial, and there is almost none about cones, despite many decades of intensive study. Fortunately, the same biochemistry, if not the same types of cells, are at work in the eyes of most other species.

When photons cause chemical changes in a series of molecules, the result after about 1 millisecond is the emission of nerve pulses along the fiber or axon, leaving the vision cell and passing upward to another layer of the retina. The vision cells work backwards from what might be expected of sensors. They allow electric current to flow (dark current) unless light strikes the cell. Then a pulse occurs which consists of the interruption of current. Nerve pulses, whether positive or negative, are somewhat erratic bursts of current and voltage that occur in milliseconds. The information is carried mainly in the rate of the pulses, with greater stimulus resulting (usually but not always) in more frequent pulses. If there is zero light, a few pulses are still sent to the brain. It is amazing that the nervous system and the brain can make so much sense out of such erratic tiny signals with no rigid relation to stimuli and no rigorous modulation of amplitude and frequency. More remarkable is the amount of noise that seems to be in the visual system, yet you see images much more clearly than the best film or video screen can provide.

One aspect to notice about the nerve connections at the surface of the retina is that there is no direct correspondence between optic nerve cells and vision cells. Several layers of other nerve cells (bipolar, ganglion, horizontal, and amacrine) are intermediate and have the opportunity to process visual signals before they pass out of the eye along the dense bundle of fibers that constitute the optic nerve bundle. The horizontal and amacrine cells pass visual information laterally to other bipolar and ganglion cells. Some kinds of amacrine cells have been found to make contact with up to a 1,000 different ganglion cells spread widely over the retina. The fovea and macula have the greatest density of output nerve fibers, corresponding to their greater resolution. There must be a blind spot on the retina where about a million optic nerve fibers gather together and exit through the back of the retina. A way to perceive the blind spot is to stare at one point about 25 cm away with one eye closed and to move a small object to the side until it momentarily seems to vanish.

The optic nerve bundle from each eye travels toward the brain (Fig. 16-2) and crosses at the optic chiasma so that signals from the right half of each eye go to the right half of the brain and signals from the left half of each eye go to the left half of the brain. The result is crossover of visual information because images of objects on the right side of the visual field are formed on the left half of the retina due to inversion by the lens. Thus objects on the right are processed by the left brain, and vice versa. The images from left and right eyes are not identical, partly because of different points of view and partly because the two visual fields do not fully overlap. However, unless one deliberately misfocuses the eyes, the two images tend to seem superimposed. The optic nerves come to a major connection region called the lateral geniculate body on each side. The left and right sides of the visual field are finally received somewhere toward the rear of the brain in the *visual cortex* (area striata) and superior collicus. As will be seen, progress has been made in locating aspects of the retinal image in corresponding locations in the brains of animals.

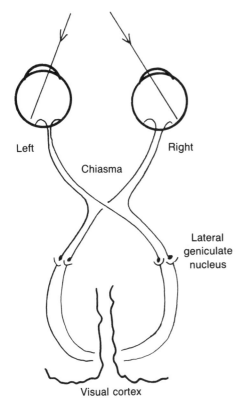

Fig. 16-2. Top view of the relation of left and right eyes to the brain, with two nerve paths traced for each side.

16.2 VISUAL SENSITIVITY AND PERCEPTION

The rod cells are sensitive to as little as one photon each, and their larger size and greater volume of material contribute to the sensitivity. They have chemical amplification systems that amplify about 10^5 times, although the details of this are yet unknown. Rods do respond four times slower than cones. The rod cells are connected so that a photon must be received by each of two cells before a definite signal is generated, so as to avoid seeing flashes of light due to occasional thermal noise in the cells. At night there is only enough light to trigger rod cells, which are sensitive from blue to orange but most sensitive to green around 500 nanometers. Figure 16-3 compares rod and cone sensitivity S (cone sensitivity here is for three colors combined).

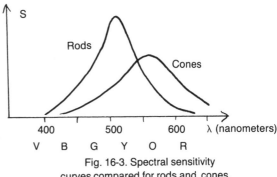

Fig. 16-3. Spectral sensitivity curves compared for rods and cones.

Colors are not detected by cone cells at night. When the light level is sufficiently high, cones take over and are most sensitive to yellow-green light at about 550 nanometers, less sensitive to blue and more sensitive to red. The shift in sensitivity as intensity increases has little to do with color vision and is called the *Purkinje shift*. While rods require about half an hour to adapt, cone cells adapt in a few minutes to the prevailing level of light. For either kind of vision cell, as the light intensity lowers, the sensitivity increases. The pupil helps out by opening to admit more light, too. More remarkably, parts of the retina adjust in sensitivity rapidly—both chemically and neurologically—so that you can see clearly a scene with both deep shade and bright sun.

Adaptation to darkness also occurs in the interaction of the nerve cells that process visual signals. In dim light, more cells cooperate to detect light, with the result that resolution or acuity is lost. Instead of two adjacent vision cells being able to detect a difference in intensity, large numbers are required to detect variation in the low-light level. Adaptation to bright light is necessary because the flood of photons overloads the rod cells. This fatigue or saturation is also a warning sign that the sensory cells are overstimulated and permanent damage might result. Adding more photons, even by a factor of 10, causes no further increase in signal when a vision cell cannot respond to more. The rate of nerve pulses from the cells also reaches a limit.

Light intensity (technically, illuminance) is a physical variable. Perceived brightness is a psychological variable that cannot be measured with instru-

185

ments. Its properties can be found by asking people to report on the results of comparing various brightnesses perceived for known intensities. Animals can also indicate simple perceptions after training. Brightness is found to be a function of the reflectance of a surface and the amount of light (illuminance) falling on the surface as well as many factors pertaining to details of the image, one's past history and knowledge, mood, and so forth.

The general mathematical rule for perception has been found to be logarithmic in nature. In simple terms, this means that large changes in a stimulus such as light intensity causes modest changes in nerve pulse rate. The rate of pulses, other factors being equal, is proportional to the logarithm (log) of the intensity. The graph in Fig. 16-4A divides the scale equally for powers of ten in arbitrary intensity units—for example, $log\ (10)^7 = 7$. For each increase in light intensity by a factor of 10, the nerve pulse rate increases by a small amount. Nerve pulse rate can vary at most by a factor of about 100. Logarithmic sensitivity allows for a sensory system to accommodate the huge variations in intensity found in nature. (It also applies in part to photographic film sensitivity.) The practical range for perception is from dim starlight (about one photon per rod cell, or an intensity of about 10^{-10} W per square meter) to about one-tenth of full sunlight (about 100 W per square meter). The total range is about a trillion or 10^{12} (about the same as in hearing!).

You do not perceive such an enormous variation in intensity. The jump from a room at night so dark that false images appear on the retina to full sunlight does not seem like a trillion-fold change. A measure of perceived brightness based on a log scale does seem appropriate. On the log scale (base 10) this range of brightness results in just twelve equal steps in magnitude, one step for each power of 10. Ten steps much compressed in relative brightness are indicated in Fig. 16-4B on a grayscale with arbitrary brightness levels. Due to paper and ink limitations, the brightest and dimmest steps cannot be shown. We are not aware of changes in the way the eye adjusts to the full brightness range, but several mechanisms are at work in overlapping ranges. They include pupil size, vision cell sensitivity, changes in the number of cells used, and rearrangements in nerve cell processing.

Another general law of perception is Weber's law, which states that the smallest detectable change in a stimulus is proportional to the general background level of the same stimulus. This can be applied to light intensity and many other variables. It is a mathematical way of saying that sensitivity decreases as intensity increases so that the sensory system is never overloaded. The law must fail at very low intensities because a sensory cell cannot become more and more sensitive indefinitely. Instead a background noise level is reached, and the sensitivity increases until the noise dominates.

Weber's law is relevant to the perception of contrast. Contrast can be scaled in various ways for both the eye and photographic film on a grayscale (recall Fig. 16-4B). In low-contrast situations, a wide range of intensity is encompassed in, say, 10 steps on the grayscale, but features may be difficult to distinguish. In high-contrast situations, a narrow range of intensity is interpreted with a grayscale, and subtle features stand out. The retinal processing attempts to adjust eye sensitivity to contrast (in accordance with the average light level and the range of intensities received) so that far more than 10 grayscale steps can be discriminated. Zero brightness or pitch black is not attainable because the darkest materials reflect a few percent of incident light. Similarly, the best white does not reflect more than about 90% of the light incident on it.

Color vision is reserved primarily for the next chapter, and only brief background is given here. The psychological variable that describes the effects of spectral color is called *hue*. The one that tells how spectrally pure the color is (distinguishing, for example, bright yellow from greyish yellow) is *purity* in physical terms and *saturation* in perceptual terms. The reader should be careful not to confuse this meaning of saturation with the earlier concept of fatigue. The approximate range of spectral colors seen is from violet near 350 nanometers to red near 700 nanometers. Just a factor of two variations in wavelength accounts for the rich colorfulness of the perceived world! It will be seen later that each of these

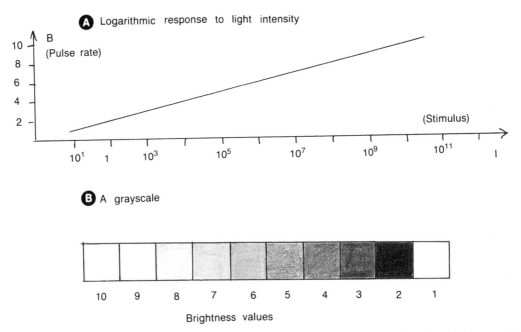

Fig. 16-4. Converting trillion-fold variation in light intensity I to a much smaller range of perceived brightness (B). The response as nerve pulse rate occurs through logarithmic sensitivity. The grayscale is a compressed brightness scale.

variables can be saturated (in the sense of overloaded). And as with intensity, there is a smallest detectable change in hue and purity. For the present, the brightness of parts of images can be discussed without regard to color, as if all the world were viewed in shades of gray.

The resolution of the eye should be limited by the graininess of the cone cells in the fovea and the quality of the lens. In fact, experiments show that the lens presents a better image than the vision cells can detect. It is possible to reverse the optics and see retinal details through the natural eye lens! In terms of angular resolution, the limit is a few minutes of arc. It can be tested by using close pairs of lines. It can also be determined by measuring the highest spatial frequency that can be seen, about 60 cycles per degree (to be discussed further). The explanation for how the eye achieves a fine focus is that feedback signals from the fovea during growth control the growth so that the shape of the eye adjusts itself for best focus.

The eye can detect changes as rapid as occurring in 1/50 second in good light. Images seem to persist until the changes are received by the brain. This is very important for seeing motion as a smooth change rather than a series of jerks. Because the speed of response is slower in dimmer light, an unusual effect can be seen by watching a pendulum swing sideways with one eye covered with a dark filter. This Pulfrich pendulum is seen to swing in an elliptical orbit because of the delay in one visual signal. The perception of motion is discussed more in a later section.

When a set of visual cells is overloaded, either by staring at a bright light or just staring at any prominent object, an afterimage is produced. This can be seen when the eye is closed immediately afterward (ignoring the colors for now). The afterimage is seen negative, with dark where light was, and vice versa. You can easily prove to yourself that this image is imprinted on the retina for a few seconds or more. It also cannot be further altered by the visual system. Positive afterimages are also possible.

The retina is sensitive to damage by intense light (less than 1000 watts per square meter, the intensity of the sun), with more energetic colors being more quickly damaging. Ultraviolet is by far the most dangerous. To some degree, the eye is protected from ultraviolet by the fact that the lens will not focus ultraviolet well on the retina. The lens does focus near infrared and is liable to damage from that source.

Some experimenters have viewed the retina as part of an optical transform from image to brain. When each cone cell is assumed to be a pinhole, the diffraction pattern of the retina can be measured or calculated. Since the pattern is a Fourier transform of the retina, it indicates the spectrum of spatial frequencies which the retina can detect. Because the cone cells are randomly located (favoring the macula), there is a noise imposed on the transform, but interference is avoided. The array of cones turns out to be neither too regular nor too random and thus is arranged for optimum vision.

The retina is sensitive to some stimuli quite different from light. Ever since astronauts reported flashes in their eyes—soon traced to cosmic rays (high-speed atomic particles) passing through—scientists have been interested in this novel but hazardous way of detecting particles.

16.3 NEURAL CIRCUITRY IN THE RETINA

Many experiments show that the retina is arranged to detect contrast in special ways. Much of the detection is "wired" into the way nerve cells are connected. Neurology or the study of nerve cells is a vast field. It will be sufficient to know here that pulses enter a nerve cell along a number of fibers called *dendrites*. The cell does a sort of chemical-electrical computation and generates a series of new pulses related in part to the inputs. The output pulses travel along a thicker fiber, the *axon*, which may be very long. The axon may branch near its end to form more connections to other nerve cells. The connections are called synapses and are one-way connections from axon to the next dendrite(s). The pulses are a combination of chemical and electrical signal and travel fairly slowly along nerve fibers. The fastest possible pulse rate is about a thousand per second (1 kHz).

The process of detecting changes in brightness at edges or spots automatically occurs in the retina by means of the way nerve cells are connected in neural circuits. The detection is done point by point over regions of the retina, so the response of vision cells to spots of light requires consideration. It turns out that such *feature detection* processes are "wired" in. Some other processing circuits have been determined, but full understanding of vision will be a long time coming. Some of the features in images for which circuitry is being identified are shown in Fig. 16-5: edges or lines (straight and curved), lines meeting at an angle (vertex), ends of lines, uniformly spaced bars, spots, and the locations, sizes, rates of movement, and orientations of these.

In the *center-surround* model, a spot of bright light on one or a few vision cells results in a signal of light for that region, modified by whatever is happening to the vision cells surrounding the central cell (Fig. 16-6). The light stimulus results in both positive and negative signals. (Here positive and negative are used loosely and not in a precise electrical manner.) Pulses from the vision cell have a positive effect when they cross a synapse whose chemistry causes stimulation of the next cell to produce pulses. Pulses have a negative effect when they reach a synapse where the chemistry causes an inhibitory sig-

Fig. 16-5. Some of the features that are or might be directly detected by neural circuitry in the retina. Position, motion, and orientation of features are also detected at some level.

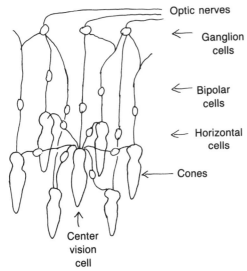

Optic nerves

← Ganglion cells

← Bipolar cells

← Horizontal cells

←── Cones

↑
Center
vision
cell

Fig. 16-6. A central vision cell and its antagonistic surround of horizontal cells. Amacrine cells are not shown for clarity, and neural connections are simplified here. Light enters from above.

nal sent to the next cell to reduce or prevent pulse generation. Neural circuitry is not very exact. The number of vision cells and bipolar cells is about the same, but each vision cell sends pulses to dendrites that lead to several bipolar cells. The "wiring" depends on the species, and the well-studied frog retina has more connections and built-in processing than primate retina.

The positive signal proceeds through bipolar and ganglion cells. The negative signal travels sideways through the horizontal cells and partly inhibits nearby bipolar cells from passing positive signals. The horizontal cells also bring in signals relevant to the steady background intensity, thus responding to the average light level of the surround. The amacrine cells (not shown in Fig. 16-6) have a dense maze of dendrites spread across the retina and bring in signals only if the surrounding light is rapidly changing. The strongest effect is from nearby vision and bipolar cells, but more distant parts of the retina have an effect, too, if large parts of the image are the same. Two outputs from the vision cell result at the ganglion cell level: an output modified by the average surrounding light level and an output modified by any rapid changes in surrounding light. The

general effect is that local vision cell response is reduced in proportion to changes in light level over larger areas, resulting in a reduction of contrast. Sensitivity to motion of small parts of the image is enhanced because a moving image is readily detected locally despite confusion from the image elsewhere on the retina.

If a tiny ring of light were presented, then the vision cells in the center would have difficulty seeing because they are inhibited. Every bipolar cell is surrounded by a ring of negatively-responding cells or antagonists. This seems to be the most basic feature-detection system, "wired" in. The purpose of this dynamic structure is to adjust the sensitivity to contrast in accord with the general light level. For other basic feature detection, there are many kinds of amacrine cells, each with a different purpose and a different synapse chemical (neurotransmitter). They are spread over the retina in different ways. One kind (*cholinergic*) is numerous with dendrites densely intermingled all over the retina and it detects direction of motion. Another kind (AII, read "A-two") is sparsely packed, with dendrites barely reaching those of its neighbors. It helps control response to widely varying light intensity. Many kinds remain to be clearly identified and studied in detail.

The detection of straight or curved edges can begin to be understood by considering how the edge crosses a region of vision cells as the edge or eye moves. Figure 16-7A shows a straight edge of a dark region that has triggered a row of cells. Information on the direction and rate of advance of the edge is generated and sent out by the horizontal and amacrine cells. That the edge is an edge is detected from which parts of vision cell surrounds are still in light. Direct interpretation of moving edges in the frog retina has long been known to result in signals sent forward to the brain, but this is still not as certain in humans.

Direct detection of the curvature of an edge is probably built into the rate at which cells are triggered as the curve sweeps across. Sprinkled over the retina are many nerve cells "wired" differently to respond to different behaviors of an edge. Various sets are triggered depending on which way the edge moves. Direction is detected by having inhibit-

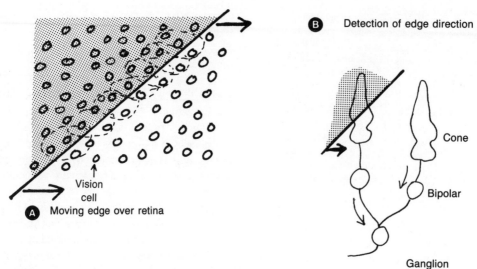

Cone

Bipolar

Vision cell

A Moving edge over retina

Ganglion

Fig. 16-7. Detecting an edge—a boundary between light and dark. Its direction of motion is detected because signals generated in front of the edge travel slower.

ing bipolar cells located to one side of the ganglion cells which they inhibit. An edge passing over vision cells seems to generate a signal which travels faster than inhibiting signals for that direction (Fig. 16-7B). An edge passing over in another direction generates inhibiting signals which reach the ganglion cells earlier.

16.4 FEATURES OF IMAGES FOR PERCEPTION

The optical image that is cast onto the retina must be distinguished from the perceived image which is made conscious to us somewhere in the brain. Just what do you see in a scene that has no familiar objects with names? Light reaches the eye after reflecting from assorted surfaces. In most cases, the surfaces absorb part of the light so that darker or colored surfaces are seen. The light may be reflected from various depths of transparent and translucent materials. (Color effects are reserved for the next chapter.) Most surfaces also have texture—that is, recognizable characteristic detail that can be identified as such regardless of curves and shadows.

For the moment, assume that the surfaces of interest are all normal to the line of vision from eye to surface. Let there be just one surface with very

simple texture or features. Consider a flat sheet of paper with just one feature, a boundary (edge) between light and dark (Fig. 16-8A). This *step change* in brightness is one of the simplest features that the eye can detect. To understand better the effect of this feature, consider a more complex image consisting of dark bars separated by light bars, a coarse grating as in Fig. 16-8B. In this case, the student of optics should recognize a spatial frequency. The regularly spaced bars have a spacing which is the spatial period, and the frequency is the reciprocal of that period (measured in inverse meters or cycles per degree). With this in mind, the single edge is recognized to have bands of spatial frequencies, whereas a grating has one basic frequency and a series of harmonics that are multiples of that frequency. In one sense the grating is a simpler feature.

The resolving ability of the eye is often tested with gratings—parallel-ruled lines of specified spatial frequency in cycles per degree—the number of lines in one degree of angular view. The lines may have sharp edges (abrupt shifts from dark to light) or fuzzy edges. The most smoothly varying lines have a sinusoidal profile in brightness (recall Fig. 14-6A). The evidence indicates that the eye/brain is more sensitive to an image with regular lines (a grating) than to a single line (a bar). This indicates

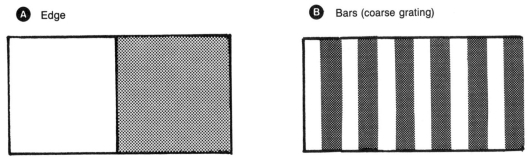

Fig. 16-8. An edge and parallel bars (with spatial frequency).

that spatial frequency is a fundamental part of perception. The breakup of a picture in two dimensions into blocks (crossed lines or bars) is discussed later.

A different sort of pattern demonstrates a different aspect of perception. Gaze steadily at the field of uniformly spaced dots in Fig. 16-9. Different patterns of dots seem to form and dissolve. The eye/brain is continually trying out fresh groupings of these minimal clues, searching for some sort of sensible organization. The active mind refuses to see just what is there, an orderly set of dots! Percival Lowell, when observing Mars, saw in a few random spots on its surface a network of straight canals across the surface. The first attempts by the brain are to see rows or columns of dots. This preserves the simplest spatial frequencies. You will probably notice a switching back and forth from the row interpretation to the column interpretation. Then groupings into squares may occur. The diffraction pattern and therefore the spatial frequency map for

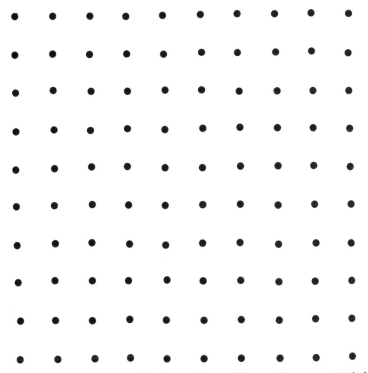

Fig. 16-9. A grid of dots for exploring the tendency of the eye to see rows and clusters.

ordered dots is a series of spatial frequencies in both dimensions, in integral multiples of the lowest. One senses rows, then double rows, then triple, and so forth. "Op" artists have exploited spatial frequency detection and uncertainties in visual processing when producing "simple" paintings consisting of ordered dots and stripes. Moiré patterns affect the visual system in similar ways.

Contrast is measured with spatial frequencies by measuring how well each frequency is seen (using perhaps a grating with variable spacing). The relation of brightness to spatial frequency constitutes a contrast transfer function. The sharper an edge, the better seen are high spatial frequencies associated with the edge and the better the contrast at the edge. Conversely, increasing the contrast raises the spatial frequency that can be detected. The human limit is about 60 lines or cycles per degree.

The detection of edges and the spatial frequencies of content clarify to the concept of contrast. To see an edge, the brightness must change abruptly as the edge is crossed. The eye seems to be able to distinguish many levels of brightness—perhaps as many as 100—within a given scene. The variations in brightness occur usually on surfaces. Accurate distinctions are possible only between adjacent surfaces. The brightness of a given surface is always determined by the brightnesses of adjacent surfaces. (Adjacent groups of retinal cells are processing the image.) The detection of brightness levels actually depends upon an edge—a boundary where brightness changes noticeably—moving over vision cells and triggering them. The eye judges brightness only from relative changes in brightness as the eye scans a scene. We perceive the boundary and not the adjoining regions. That brightness is relative and subjective can be seen in patterns on white paper which cause certain areas to appear whiter than white and others blacker than black. Sunspots are thought of as dark and they appear and photograph that way, but they are almost as brilliantly white as the rest of the sun's surface.

A closer study of the perception of contrast reveals how there are many levels of inhibition in the retina. The inhibiting signal for a bright spot surrounded by light is itself inhibited if a much larger area is illuminated. The result is that midrange spatial frequencies (moderate detail) are more easily seen than high frequencies (tiny detail) and low frequencies (large uniform areas of light or dark). Moreover, at an edge contrast is enhanced, with the light side seeming lighter and the dark side darker, in narrow bands or fluting near the edge. Examine the edges in the grayscale of Fig. 16-4B and consider the brightness contour shown for an edge in Fig. 16-10A.

In the Craik-O'Brien effect, a boundary of both light and dark is constructed on a uniformly light gray surface according to the contour shown in Fig. 16-10B). The eye is trapped into assigning lighter to one side and darker to the other. If the boundary is gradated darker at the edge of a gray region, the region will appear lighter than the adjacent region even though the adjacent region actually is lighter as indicated by the contour in Fig. 16-10C. The brightness variations at a boundary are assumed by the eye/brain system to apply at least partially to large adjacent regions. This is a result of the tendency of the eye to examine only boundaries where there is contrast and to be unable to perceive accurately large featureless areas.

Contrast in most scenes depends on the type of illumination. Direct illumination gives sharp shadows and widely contrasting brightnesses. Indirect illumination reflected from objects gives less distinct shadows and slight contrast. The eye/brain system continuously attempts to ascertain the source of illumination from clues in the scene so that it knows which interpretation to apply. The strongest clue is the relative brightness at boundaries. Look back at Fig. 16-8A and decide if what is shown is a boundary between light and dark on flat paper or a flat image of a corner. The interpretation could be an edge uniformly illuminated or the outside corner of two identical walls, one in shadow and one well-illuminated.

By its nature, transparency is almost invisible (except for reflections). The eye/brain system has a mechanism for supposing transparency as part of the visual world. The transparent material need not be colorless. Where a lighter region seems to over-

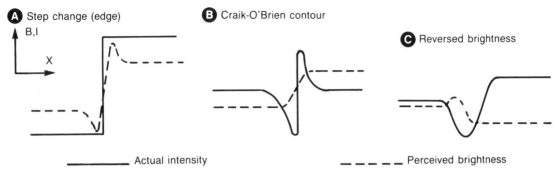

Fig. 16-10. Contour graphs of actual illuminance (intensity I) and perceived brightness B at three kinds of boundaries.

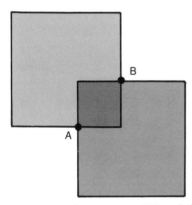

Fig. 16-11. The illusion of transparency (if the relative brightnesses are correct).

lap a darker region (Fig. 16-11), the lighter one is assumed transparent (if the geometric relationship is sufficiently simple or obvious). The boundary of the supposed underlying region must be predictable where it is covered. At the least it must be continuous at the two points A and B shown, where the boundaries intersect. The brightness of the supposed transparent material must have a special relation to the underlying material in order to see transparency instead of three connected squares. The required brightnesses can be calculated (see article by Metelli).

The way texture is perceived is directly related to the associative connections among nearby vision cells. Texture is first seen as clusters of spots of equal brightness. When looking at texture, the brain seems to search for order. Order is found by finding clusters of identical brightness. If these clusters repeat in any regular way, the texture has a basic unit that is readily detected. Relations among the units are then sought. Random dots and other random pictures using simple elements are valuable research tools for understanding how we perceive texture. Bela Julesz has pioneered this ingenious areas of study. For example, in a field consisting only of the letter "R" repeated many times and randomly oriented, the mirror images of R are impossible to distinguish at a glance from the normal Rs. But if the Rs are oriented the same and arranged in rows and columns, a group of reversed Rs is easily spotted in a field of normal Rs.

The texture of a surface may give clues as to the orientation of the surface. If the details of the texture appear smaller and closer together on certain parts of the surface (Fig. 16-12A), then the mind has learned to interpret this as a surface which varies in distance, with some parts closer and other parts—those with the finer texture—more distant. This is another way of saying that the surface is not normal to the line of vision but rather tilted at an angle as in Fig. 16-12B. Texture also has an effect on the way light is reflected toward the viewer, tending to absorb and diffuse it more. Heavy texture results in a texture of shadows (for example, tree bark).

On a larger scale, surfaces adjoin each other at edges—for example, the corners of a box. These edges are the features most readily tracked and recognized by the eye and give the strongest clues to identifying an object. Different surfaces receive different illumination. Even if the object is all one

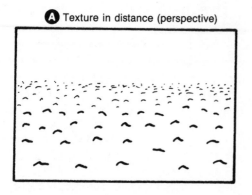
A Texture in distance (perspective)

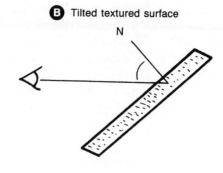
B Tilted textured surface

Fig. 16-12. Diminishing texture indicates a tilted surface seen in perspective.

intrinsic color (for example, a box), the various surfaces can vary widely in brightness. If the object is recognizable, the visual system assumes (sometimes without justification) that all sides are the same and that the differences in brightness indicate which way the incident light is shining on it.

Once the visual system has determined the way distance is shown via texture, it can judge the relative sizes of objects regardless of distance. Conversely, if objects of known size such as people are scattered about—some near, some far—the depth of field is immediately clear. When known objects are in proper relation to known background texture (for example, trees in a forest), then the illusion of depth is improved. The elements of the texture need not be identical nor arranged in a regular manner such as on a grid. Only the average size and spacing need diminish properly with apparent distance (recall Fig. 16-12A). In the distance, the visual system expects to see an ever wider field of view and therefore more elements.

16.5 MOVING IMAGES

Whether the object moves or not, the image on the retina is almost always in motion. The natural behavior of the eye is to scan scenes in *saccades* (Fig. 16-13). The eye rotates very rapidly from one point to another, pausing to fixate on certain points. The scan can take place in any direction and occupies only about 10% of the looking time. The saccade is designed mechanically to occur as rapidly as possi-

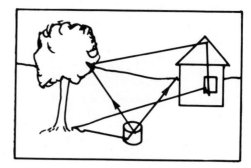
Fig. 16-13. Example of part of the scanning path (saccades) of an eye.

ble and tends to overshoot. It is then corrected by a short glissade. The scan consists of separate horizontal and vertical components of motion. Experimenters have found that the focus of vision jumps almost randomly from one point to another—vertically, diagonally, horizontally. The jumps vary greatly in length and depend on the content of the scene. A small complex region receives many small scans and is often returned to. Edges and angles receive much attention. The saccades bring small regions of the image to the fovea where they may be examined in fine detail. Special features that "catch" the eye are returned to often, for fractions of a second. These features may be prominent for their contrast and curvature, or for their relation to known objects of emotional interest. For reading print the eye learns to scan a line at a time in a few quick jumps.

Despite the flicks of the eye and the consequent erratic movement of the image on the retina, a person perceives a stationary scene, or at least thinks he/she does. This is a clue that much processing of the image is done at and after the retina before the results are made conscious. When the object to be studied is itself in motion, then the eye may track it but also continues scanning in saccades. Despite the double motion, people see moving objects as if they were stationary. When there is no background, the eye and/or head can still track a moving object, hold its image approximately stationary on the retina, and accurately estimate its motion. Information from the retina and from the eye muscles seems to compensate accurately and keep a perceived image fixed. In one application, this compensation is a disadvantage. During the driving of a vehicle, one's tendency not to see the motion of passing objects has been shown to add about 0.3 seconds to the reaction time needed to see and avoid a collision.

Closer consideration of how eye and image movements are distinguished gives the surprising result that the brain commands the whole process. Indeed it would not work for signals from eye muscles and retina to interact on their own, because the muscle signals arrive after the retinal signals. The brain both commands the eye or head to track an object and sends a delayed signal to the retina to cancel the image movement. A complex image sliding around on the retina can be altered directly in the retina circuitry to perform a different specified motion. You can test this quite simply by relaxing the eyes and gently rotating an eyeball with a finger to the side. The image seems to move the wrong way, contrary to eye rotation! The brain has commanded the image to move to compensate for eye motion. You can also test that an afterimage does not move under the same conditions.

Motion itself has persistence. If one watches a spinning object, then stops it, the object seems to spin the other way for a short while. If one watches water falling for a while, then looks at a stationary region, the region will seem to move upward. The apt name is thus the waterfall effect. An implication is that there must be feature signals for rotation clockwise and counterclockwise. While there is some

evidence that this effect is produced in the brain, it seems more likely it is produced in the retinal nerve system. Every kind of signal from the eye seems susceptible to a fatigue, and excessive repetition leads to an opposing signal.

Persistence is evident every time we watch movies or television. A series of still pictures are presented, but we see continuous motion. The pictures are changed more rapidly than the eye can follow, but the small changes from picture to picture are perceived as smooth motions. The picture rate must be faster than about 50 per second to cause the illusion of motion, otherwise a flicker is detectable. Each movie frame is shuttered open and shut three times before the next frame is presented at the rate of 24 per second. Television pictures are changed 60 times per second. Electric lights flicker 120 times per second but seem steady.

Bright flickering light at a slower rate such as 5 to 10 bursts per second seems to interfere strongly with the nervous system, starting with signals from the retina. Strobe lights are exploited in this realm for entertainment, to agitate or otherwise confuse people. Some are susceptible to epileptic seizures, easily triggered by certain rates of flicker.

There is more to the perception of motion. The brain seems to fill in intermediate positions between two pictures of an object in different positions. If two lights are flashed in an alternating pattern such as at a railroad crossing signal, the light appears to be a single one moving back and forth. Whether or how this works is uncertain. There seems to be a general movement program for the eye-brain-retina system that assumes intermediate positions and tells the eye to scan between two known positions to find them. Some evidence indicates that the brain assumes a simple three-dimensional shape for a barely glimpsed object. The supposition of intermediate positions has primordial survival value. An animal stalking another passing behind trees must track the prey by matching a glimpse in one place with a glimpse in another. Features must be matched. Experiments show that the features matched first are brightness levels. Other rules seem to be applied by the brain to track moving images seen intermittently.

There is still more. The eyes have been found

to concentrate on features of a moving scene of which the person is not consciously aware. There is retinal and mental processing of information never brought to consciousness. While driving along a street, one's intention may be to observe those aspects of the scene pertinent to safe driving, but flashing lights, strongly contrasting objects, or an attractive pedestrian will divert visual attention before there is conscious recognition of the diversion and its consequences. The proposition by artists that they arrange features in a picture to attract the eye to move certain ways has been confirmed by experiments, with some margin of error for the subjective aspects of such studies. The study of fixation in motion, or of saccades, has other practical applications such as enabling efficient reading of road signs or study of technical photographs.

The earliest work showing feature analysis occurring in the eye itself was done by Jerome Lettvin and others in the early 1960s, working with frogs. They found output nerve fibers dedicated to various kinds of different visual information including edges and general light levels. More importantly, the edges and other features had to be moving to be perceived by the frog. Eyes are most sensitive to moving edges near the limits of the retina—about 100° from center in humans. This surely has survival value. The identity of an object cannot be determined at the limits of the field of view, and the immediate response is to rotate the eye/head to bring the object to the center for scrutiny.

Various studies of the perception of movement have revealed how the eye/brain system is naturally adept at performing some of the operations of calculus, regardless of the mathematical education of the person. One experiment showed motion through a slit so that the whole moving object could not be seen at once, just the location and rate of change of its motion. As with all aspects of vision, perception of movement has many more aspects than can be covered here or are well understood. The discussion has reached the outer bounds of optics, and the reader is left to references on the psychology of perception.

16.6 STEREOPSIS AND SEEING DEPTH

Many clues in images give an indication of absolute and relative sizes of objects seen. Familiarity with the objects is usually necessary, or else the eye is easily fooled. If sizes are known, then the apparent sizes and the relative sizes in the image are clues to the distances of the objects. Known or regular objects must diminish in size and draw closer in spacing at a certain rate with distance. Texture indicates depth because the texture grows smaller in the distance. In scanning a scene, each eye changes focus rapidly in an action called a *vergence* (the counterpart of a saccade). The focus is rapidly shifted among near and far objects, testing out the depth of the scene. The amount of vergence needed is signaled to the brain. More accurately, the brain sends focusing signals to the eye so that it can more rapidly compare the resulting image with the depth of focus. How the eye-brain system knows an image is in focus sounds obvious yet is a difficult problem.

Other clues to depth include presence of shadows, increase or decrease of overall brightness, lighter and less-saturated color, superposition of one object outline over another behind it, reflection of light from one object onto another, indistinctness of details in the distance, location of small features near a horizon or above larger features, and convergence of parallel lines toward vanishing points. A clue from parallax is that a moving observer sees foreground objects that seem to move past background objects. When an object is moving toward or away from the viewer, its location and speed must be judged by its change in size. It is vitally important to judge accurately an approaching object, so the brain assumes that growing size results from approach, not from swelling like a balloon. All these clues work with one eye alone. When some of these clues are present in flat pictures, the mind is fooled into perceiving depth.

The combination of images from two eyes provides true *stereopsis* (Greek for solid vision). Although the term binocular is often used for stereoscopic vision, it is best retained for reference merely to the use of two eyes. Stereopsis is the process of forming a *cyclopean* (one-eyed) view from two slightly different views. Eyes are located naturally in a horizontal plane, so that one is left and the other right. For high and low objects, the eyes must rotate upward or downward to bring the object into their plane for judging depth. Major information is provided by the eye muscles, which signal which directions the two eyes are pointing when focused on a particular object. This translates to a measurement of its distance by triangulation. When the object moves, tracking its distance is even easier (Fig. 16-14).

When both eyes look at the same object point, there is a disparity between features in the two images. The eyes attempt to minimize the disparity if it is not too great. Large disparity results in an inability to fuse the images, and double images (pleasantly called ''diplopia'') results. Stereoptic information is primarily in terms of disparity. The signals from each eye are compared in the visual cortex, where individual brain neurons have been identified (mainly in animals) as sensitive to various disparities.

Stereoptic perception is as accurate as retinal structure allows. The disparities in the left and right images of a distant object may be smaller than the size of a visual cell, yet the object is still judged accurately in its distance. As might be expected, those

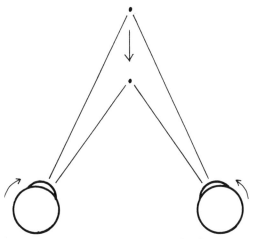

Fig. 16-14. Binocular detection of a distant object moving closer.

animals with eyes on the sides of their heads have poor binocular vision. There is little overlap of left and right images. These animals are usually preyed upon by animals with frontal eyes and better binocular vision.

One of the most powerful tools invented for the study of stereopsis and other aspects of visual perception is the random dot pattern. These are generated by computer. A screen might be filled randomly with 10,000 dots in a square pattern of 100 by 100 pixels. Each pixel is either black or white. In most experiments, two random dot pictures are presented, one to each eye in a special viewer. Less accurately, the dots are printed in red and green in the same picture and viewed through red and green filters. Each eye separately sees no order or pattern in its random dot picture. But during computation of the patterns, a special bias has been built in.

For example, a central square composed of many dots is displaced one or more dots to the right or left of the same dot pattern in the other picture. When the eyes adjust to seeing these pictures stereoptically—a feat easily learned—a square in depth is seen, either in front of or behind the background of dots, depending on which way the shift was made (Fig. 16-15). The amount of depth depends on the disparity incorporated. The perception of a simple object in such pictures bypasses all retinal processing and occurs in the brain where the two pictures are matched for disparity. It is amusing to contemplate that the retina is utterly thwarted at finding features in a random dot pattern. But each eye passes on almost exactly what it sees, leaving it to the brain to make sense of the random dot pictures.

Many more findings have been made with random dot pictures than can be summarized here. When the dot patterns are generated and presented in real time (like thousands of fireflies blinking), the time limit for stereoptic processing is found. The left and right eye images must be presented within 0.05 seconds, or there is no stereopsis. Nevertheless, the number of dots which the brain is willing to process per second is very large, larger than most

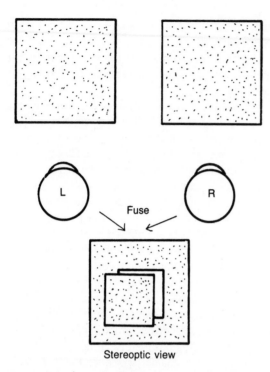

Fuse

Stereoptic view

Fig. 16-15. Random dot patterns shown to each eye result in an object shown in depth at the level of cyclopean (stereoscopic) perception when the dots have a planned disparity in their correlation.

computers can generate. The precise dot pattern received by one eye is stored for a short time to compare with the pattern received by the other eye. Varying the delay with which the same dots are seen by both eyes results in illusions of motion as well as depth. And while each pattern is identical (except for a planned disparity), the patterns need not be the same size for stereopsis.

A puzzling aspect of depth perception is depth inversion. You may have seen pictures of lunar craters that appear to be mountains and pictures of faces that appear to be molded internally like the interiors of masks. Even when these objects are viewed personally in three dimensions, the eye is tricked when the illumination comes from an unexpected direction (below). But why does not the ability of two eyes to measure the depth of features of an object result in the correct perception regardless of illumination? The present inadequate answer seems to be that objects that the brain can recog-

nize in inside-out form will be seen thus despite stereoptic clues.

As with most visual clues, the mind is flexible about which ones are accepted when perceiving three dimensions. A rectangle of wire projected as a shadow onto a flat screen appears simply as a rectangle until the actual object is rotated (Fig. 16-16). Then the shadow seems to be a growing and shrinking rectangle. The mind finds enough clues to assume it is a rectangle and visualize it rotating in three-dimensional space. Stereoptic vision would find here merely a two-dimensional shadow on a flat screen. This finding is overridden when the possibility of an object in three dimensions is realized. What process makes us prefer to see a rectangle rotating rather than shrinking and growing? Is it simply that rotation is more common than shrinking and growing wire frames, the implementation of which may be difficult to imagine or create? One aspect which cannot be ascertained on a two-dimensional screen is the direction of rotation.

Besides directing a left picture to the left eye and a right picture to the right eye, at least two other ways to control stereoptic vision have been developed. The advent of computer-processed pictures has allowed the ready generation of stereo pairs, in a modern revival of 19th century fascination with stereoptic viewing. When pictures are printed side by side, a simple viewer with lenses is sufficient to fuse them. Some need only hold a card and the head in the right places. Pictures called *anaglyphs* can also be printed in red and green or other well-separated colors superimposed and viewed with corresponding color filters (one for each eye as mentioned earlier). In theaters it is not practical to look at two separate screens, so the left and right images are given different polarizations and viewed from the same reflecting screen with polarizing glasses. The quality was never satisfactory for three-dimensional movies, despite the effect.

16.7 VISUAL ILLUSIONS

Those who study vision prefer to use objective data and to trace all results to consequences of physical laws. Visual illusions, while seemingly subjective phenomena, can be interpreted carefully to infer

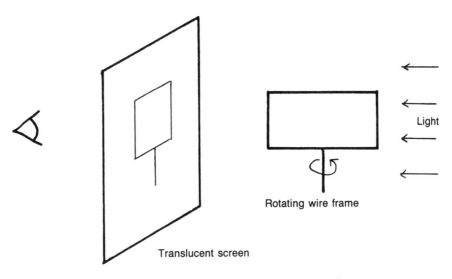

Light

Rotating wire frame

Translucent screen

Fig. 16-16. A rotating wire frame is perceived three-dimensionally as such even though the eye sees only a shrinking and growing shadow on a flat screen.

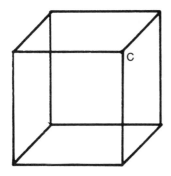

C

Fig. 16-17. The Necker wire-frame cube, a bistable figure.

many of the ways that the brain processes visual information. Most illusions occur in the brain. Only a few are traceable directly to fooling retinal processing. The retina is generally too dumb to be fooled. Some illusions reveal that the brain is very good at processing complex geometric knowledge, including properties of curved surfaces. An illusion already mentioned is the Pulfrich illusion involving the different response times of the retina in dark and light.

One of the most famous illusions is bistability. A wire frame cube (Fig. 16-17) will alternately appear with corner C outward or with C as the bottom corner of an empty box. This is the Necker cube. The brain has two choices of interpretation and so makes both consecutively. If the cube were

given perspective, then one version would likely not appear. Some artists have exploited bistability, and the references have many other examples. Another common one is the goblet that can be seen as two faces instead. This and the next illusion involve a figure and ground. When one sees an object or feature as figure, it is impossible to see the same object as background.

The two lines in Fig. 16-18A are the figure and seem identical and parallel. When a special background is provided for this Ponzo illusion (Fig. 16-18B), the two lines seem obviously nonidentical, even nonparallel. Such visual illusions have been shown to occur after the point of stereopsis in the brain. When the figure is presented to one eye and the ground to the other in a random dot setting, the illusion occurs even though neither eye alone is given the full means to see it. The Ponzo illusion shown is an abstract form of illusions based on perspective. The converging background suggests a space diminishing in the distance, so that one of the lines is perceived as more distant. If it really is the same size as the other but more distant, it appears larger.

The eye-brain system readily constructs missing lines, edges, and boundaries in order to make objects appear simple and complete. For example, the fragmented objects in Fig. 16-19 indicate angles

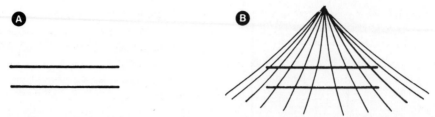

Fig. 16-18. A variation of the Ponzo illusion for two parallel lines on a special background. Curvature may also be seen.

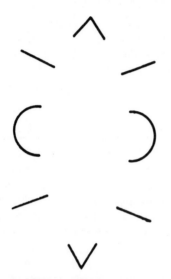

Fig. 16-19. Minimal clues are needed to see a subjective contour.

and points along the boundary of an invisible curved shape, yet no curve is shown. The brain has supplied the curves as a best match to the clues given. The curves are *subjective contours*. Another aspect of this illusion is that the brain tries to avoid incomplete lines. It chooses the interpretation that makes sense of lines that end suddenly. The background for the Ponzo illusion could have been omitted and been indicated subjectively only. The illusion remains, indicating that the brain believes background lines are there.

Negative aftereffects can be obtained in black-and-white and in color. We are perhaps most familiar with seeing an opposite color after staring for a while at a vivid color. But aftereffects occur with just about every perceived feature. Motion one way results in illusory motion the other way. Lines tilted one way are then seen tilted the other way. Curves

one way then appear curved the other way. For some effects, one must stare more than a minute before visual processing locks onto one mode and fatigues. Aftereffects occur because the eye-brain system is attuned to detecting change and detuned to seeing the same thing for more than a few seconds.

A notorious illusion is the "impossible figure." An object is drawn in sufficient detail that it seems to be a representation of a real three-dimensional object. Then the details are connected unconventionally. This sort of figure was exploited by the artist M. C. Escher on an ambitious scale.

A form of experimentation that changes illusion to reality is to have a person wear mirror or prism arrangements that present the world upside down, reversed right and left, or with wide eye separation. Most people readily adjust to these frustrating changes in a few days, then adjust back when the apparatus is gone.

16.8 HIGHER IMAGE PROCESSING

The physical image lies on the retina. What is it that we see in our minds, if we do not see the retinal image directly? Is there a copy of the retinal image put on another organ deeper in the brain? The simple answer is no, but there are some traceable mappings from real image to perceived image. A fuller answer must wait for the results of some of the most fascinating research in science. Only early results and speculation can be described here.

The first level of abstraction of features from the retinal image has been discussed, with edges as an example. For humans there is still some doubt as to whether information more complex than this is abstracted in the retina, then sent to the brain. The evidence indicates that a wide range of feature signals are sent forth from the retinas over the million or so nerve fibers of the optic nerves. A second level of processing of visual input is the determination of orientation and motion of features such as moving edges. The orientation of edges is distinguished to within 10°. These signals may originate in the retina.

A third level of processing seems to be the determination of relationships among the extracted features. This may be termed the abstraction of relations. A possible example is the determination of the outlines of objects from edges. The levels of abstraction continue—how many is not known, nor just what occurs at each level. Artists are unheralded researchers in perception who exploit complex aspects of visual processing. The techniques are many. Where the eye looks and scans, a picture can be controlled by arranging prominent edges on selected subjects and by using contrast to make certain regions appear lighter or darker than shown. The artist might also repeat shapes and outlines, exploit one or more geometric illusions, and provide clues to depth.

The neurons involved come in many types and shapes. How they work together is poorly understood and beyond the scope of an optics book. The visual cortex seems to have about six layers of neurons with distinct functions. It is sometimes called the striate cortex because of the layering. Each layer may have sublayers; for example, there are sublayers responding to edges with certain orientations. The total processing area of the visual cortex seems to be at least 10^5 times larger than the retinal area, and just as fine grained. The locations of features in the image are mapped to corresponding locations in layers of the cortex, but the mapping is chopped up so that orientations and left-right information are included. The article by Livingstone and Hubel shows the known extent of some processing channels. It is clear that most visual information is in terms of brightness that is processed in the "magno" system separately from color information carried in the "parvo" system. This seems to continue up to a high level of abstraction.

At some level in the brain, spatial frequencies are detected and processed. Spatial frequencies have been shown to represent the sizes of features and to be more accurately seen than edges and bars. The conversion of image features to spatial frequencies involves a Fourier transform, but no mechanism for doing the transform has been identified. Two dimensions of spatial frequencies are necessary, as shown by considering square grids. The illusion of diagonal white lines seen on a square grid (crossed gratings) represents spatial frequencies for lines at 45° angles. Since the transform of a grid of lines is a grid of dots, the appropriate spatial frequencies are represented by the dots along diagonals as well and may be so generated in the brain. However, the brain prefers the horizontal-vertical orientation for image analysis.

The recognition of faces is an area of fascinating research because we can learn to distinguish so many on the basis of so few features. It has been shown that only about six features (eye spacing, hair, etc) must be compared to distinguish about 1,000 faces. At the fundamental level, a face provides good subject matter for two-dimensional spatial frequency analysis. Typically a photograph is converted into a block image, where each one of several hundred square blocks (simplified in Fig 16-20) has uniform brightness over the area of the block. This brightness represents the average brightness of the original image in that position. Commonly seen is a portrait of Abraham Lincoln or other historic person altered in this way. When seen at a distance or by squinting, the block face appears almost normal and recognizable.

The breakdown of the image into blocks provides a lowest spatial frequency for the information, the spatial frequency corresponding to the block size. Such a block image has much pictorial noise, however. High spatial frequencies due to the sharp edges of the blocks carry information irrelevant to the face. Whether this noise is ordered in the array of blocks or spread randomly over the picture makes little difference in the way it hides information. One type of experiment has found that the face is most recognizable if spatial frequencies in a critical band only two to three times higher than the block frequency are filtered out, preserving some sharpness (some amount of edge). Filtering out all high spatial frequencies actually makes the image too blurred to recognize. Such spatial frequency experiments determine what aspects of a face or any other object play the most important roles in recognizing it.

A symmetry is imposed on perceived pictures

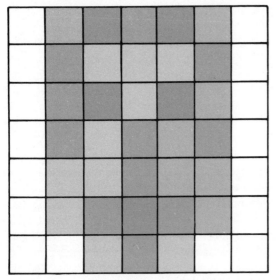

Fig. 16-20. A simplified blocked picture of a face, illustrating two dimensions of spatial frequencies and the noise introduced by the high frequencies of the edges.

by virtue of the direction of gravity—vertically downward. The eye-brain system tends to assume symmetry in this regard unless signals that the head is tilted or upside-down are sent to the brain. The symmetry is one of equating left and right. Recognition has been shown to be worst when the head (or object) is rotated to upside-down.

The brain is accustomed to working in a three-dimensional world. While each retinal image is two-dimensional, the interpretation favors three dimensions even when stereopsis is lacking. Each retinal image is a perspective view of the real world, transformed from three dimensions to two. The brain must restore three. Research has shown that when the object is as simple as an orbiting bright light in the dark, the mind interprets it as orbiting circularly in some tilted plane rather than moving elliptically on an imaginary screen (Fig. 16-21). The element of motion is required, however.

For further processing, the brain seems to have a short term or iconic memory of the partly processed retinal image. The storage time depends on the spatial frequencies stored and varies from 0.3 to 0.5 seconds. An image can be put into iconic storage with a flash exposure on the retina as short as 0.05 second. At some later stage, the brain forms

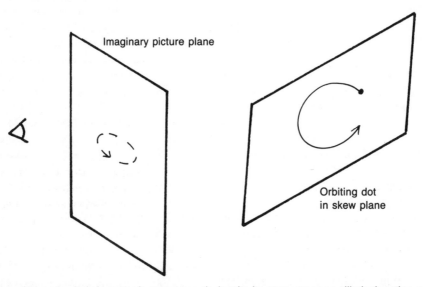

Fig. 16-21. A dot orbiting circularly in space is seen as such despite its appearance as elliptical motion on an imaginary screen in perspective.

a mental image of the perceived scene. This mental image seems to affect lower levels of processing, all the way down to feature extraction so as to render processing more efficient. In some sense, the brain finds what it is looking for. Longer term retention of pictures in memory varies widely in people. Rather than being stored like photographic plates, or even as holograms, what seems stored are identified features in the original image. When a picture is to be recalled, it is probably constructed from fragments of information about its content, using general knowledge of what ought to be (but might not have been!) in the picture. What is the nature of imagined images—ones constructed from visual elements never seen as a whole before? This is the point at which the discussion definitely departs from present science.

The eye-brain system uses parallel processing as much as possible and is a predecessor of the technological move toward parallel optical and computer processing. The evidence on parallel processing is that the brain can locate a class of obvious features in a scene very rapidly but takes much longer if a single unique feature is to be found (using serial processing). As shown by glancing at Fig. 16-22, the

Fig. 16-22. Parallel visual information processing illustrated by the ease of picking out and counting a distinct class of objects, and serial processing by the difficulty of finding a tiny defect (a broken line).

number and location of the features that are small circles are seen in an instant, whereas telling which line is broken requires examination of details of every feature.

The attempt to construct artificial vision systems run by computer for robotic applications has contributed to the understanding of human vision. The question of whether to mimic aspects of human vision must be decided. So far the brain can do many more real time calculations in image processing than a computer working at the same resolution. Aside from this limitation, almost none of the processes used by the eye-brain system has been duplicated with any reliability in an electronic system. Early attempts involve edge detection (including motion), filtering out noise and unwanted information, and enhancing contrast. Mathematically the approaches resemble center-surround processing! Among further challenges are the determination of missing or hidden lines and surfaces and the establishment of a unique interpretation of potentially ambiguous images.

16.9 OTHER EYES

Nature has exploited every optical principle in attempting visual organs. A few species (for example, the nautilus) have pinhole camera eyes, using a small aperture in place of a lens. Many insects have compound eyes, where a lens and vision cell are repeated thousands of times to form thousands of eyes working in parallel. Some creatures use optic fibers to carry images. Some do not have an imaging forming system, just the sensing layer itself right up front. Some use mirrors. Some eyes combine several of these modes of operation, such as a crab's with lens, mirror, and optic fiber. An arthropod has a scanning eye, with a moving lens scanning the image across a single sensor. Cats' and other animals' eyes that seem to glow in the dark are simply reflecting light from an existing source. There is a reflecting yellow layer beneath the retina to increase the light collected by the receptor cells.

The compound eye (Fig. 16-23) is broken into thousands of approximately hexagonal units called ommatidia. Each ommatidium has its own sensors and a narrow field of view overlapping with its neigh-

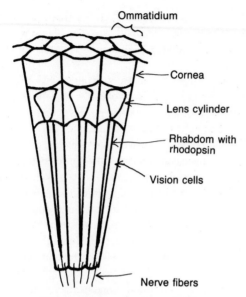

Fig. 16-23. Compound-eye structure.

bors. The size, number, and visual angle of the ommatidia vary widely according to the insect's lifestyle. The second lens of each has an unusual refractive structure. The index decreases away from the axis. Light is focused onto the rhabdom sensor (still based on rhodopsin). The sides of some ommatidia are opaque in bright light and transparent in dim light. Or adjustment is made for light level by a pigment that migrates to increase the sensor area. Vision is more blurred in dim light but also more sensitive. Some insects have ultraviolet-sensitive om-

matidia with molecules oriented in two directions to detect polarized light.

Each ommatidium with its set of vision cells has one output nerve fiber. These intermingle in a neural processing network and result in about one final output fiber per unit. The compound eye is nearly a parallel processor of visual information but no more so than the human eye. In a variant of the compound eye, each unit works to bring light rays to a common focus on a larger array of vision cells. Recently the concept of the compound eye has found technological application in an improved wide-angle lens system.

While animals have not found ways to form silvered mirrors, they have exploited interference film mirrors, using many layers of biologically formed dielectric a quarter wavelength thick to form a good reflector. The materials are alternately cytoplasm (index 1.34) and guanine crystals (index 1.83). The crustacean species *gigantocypris* has a miniature telescope with vision cells at the focus of a complexly curved concave mirror. It lives in the dark ocean and sees only luminescent creatures. The scallop has a lens corrected by a mirror that forms the actual focus on vision cells in each of its 60 or so eyes. A species of lobster has two-dimensional corner reflectors arranged in a square array in compound eyes. Some species have white reflectors with assorted film thicknesses so that many colors are similarly reflected. Others see a world colored by interference.

17

Color

This final chapter fills out the spectrum of topics on light, lasers, optics, and perception by discussing the physical and psychological bases of color. Color is understood to refer to the visible spectrum, red through violet, although wider spectral ranges are translated into "false colors" in order to display spectral information efficiently and simply. Intensity (illuminance), and its psychological correlate brightness, are an intrinsic part of color. The variables of color (perceptual correlate, hue) and purity (perceptual correlate, saturation) are also very important in this chapter. Notwithstanding these distinctions, casual discussion of color tends to use the term color in the psychological sense of hue. But saturation of hue should not be confused here with its other meaning as an overload of sensors.

Color has played many roles in earlier chapters. Optical systems involving dispersion and interference are wavelength-dependent and reveal colors of the spectrum intentionally or unintentionally. Discussion of atomic energy states has shown the sources of different colors. The atmosphere was seen to have subtle effects on color—for example,

distant snow appearing yellowish. Much more commercial technology depends on the properties of color than can be reviewed here. Excursions into the artistic applications of color are necessarily limited. Illustrating color principles with black and white is necessarily limited and requires imagination, but neither can any color printing process accurately show all possible color phenomena (a later topic).

17.1 VISUAL PERCEPTION OF COLOR

Our eyes have evolved to see those wavelengths that the sun provides. It is no accident that the strongest color emitted by the sun (yellow green, nanometers) is the color our eyes are most sensitive to. There are limits on spectral perception that begin with what wavelengths pass easily through the atmosphere and end with what elements and molecules are available that will interact with wavelengths of interest and can serve as detectors of color. In terms of light, the spectrum traditionally has its principle colors named red, orange, yellow, green, blue, and violet. Sometimes a deep blue called indigo is interposed so that the initials of the color names form the name ROY G BIV. Except for "I," these initials are used to indicate colors in black-and-white illustrations here.

The spectral range has often been referred to with a shorter list: red, green, blue, and perhaps yellow. There are no official color names or standard colors established in nature. Why our eyes tend to see the color of the spectrum in the usual named bands (blue, green, etc.) is uncertain. Why yellow occupies such a narrow band is uncertain. Pigments of the eye are sensitive in bands broader than any particular color. There is nothing special about red, yellow, green, and blue. We could just as easily choose orange, chartreuse, turqoise, and magenta as the standard colors. Someday standard colors may be defined in terms of the emitted wavelengths of assorted atoms and molecules. The spectrum provides millions of pure colors at present accuracy, distinguishable physically but not perceptually. Not all colors we can see are in the spectrum, however. Consider brown, magenta, tan, and olive.

In their cone cells, eyes have three kinds of pigment molecules (each a form of iodopsin)that serve to help detect three different bands of color, all overlapping. Each cone cell has just one of the three pig-

ments, as it must if a discriminating color signal is to be generated. The state of knowledge of biophysical color perception is good, but the molecular mechanisms in the cones is very poorly understood. The three kinds of cones are sensitive to red (peaking at 577 nanometers, really yellow orange but commonly called red), green (peaking at 540 nanometers, on the yellow side of green), and blue (peaking at 447 nanometers, near the violet side of blue). They have overlapping absorption (and sensitivity) curves, as in Fig. 17-1, with the blue response extending into violet and the "red" response extending into red. Spectral absorption is related to sensitivity because light of a particular color absorbed is light energy used to activate the iodopsin. Color discrimination is strongest in ranges where two pigments are used and where the response curves of each have substantially different slopes (rates of change as wavelength varies).

The blue cones have much lower sensitivity than the green and red. This may be due to there being many less blue cones (a few percent of the red and green count). But as the light level dims, blue vision is last to be lost, as one might expect from blue light being more energetic. Sensitivity to yellow extends the farthest to the edges of the retina (rods

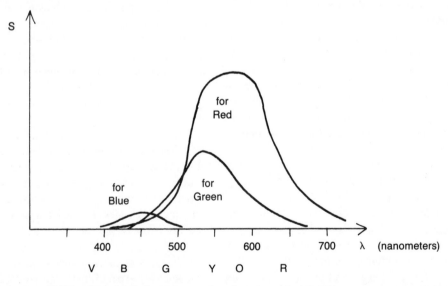

Fig. 17-1. Relative sensitivities of cone cell pigments as a function of wavelength. The red and green pigments are not most sensitive to the colors commonly used to identify them.

dominate the outer edges of vision). This can be seen by looking at orange through the corner of the eye. The red-yellow mixture in orange is seen as yellow. Yellow and blue are the most difficult to see in the distance. No bright colors remain bright in the distance, indicating that large retinal images are needed to see color vividly.

A particular color is sensed from the relative response of the three kinds of cone cells (a theory first suggested by Thomas Young). The evidence indicates that the signals from the three kinds of cones are combined in retinal neural processing before being sent to the brain, but few details of "wiring" are available yet. Center-surround connections must be present since edges and other features can be detected in color. The interconnections among the horizontal and amacrine cells serving the cones are a maze to be unraveled.

General suspicion that color sensing involves inhibition of opposing signals has resulted in the opponent model of color processing. This largely unproven theory works somewhat like color television (covered later). As shown in Fig. 17-2, signals from red, green, and blue cones are combined in the retina. Both the sum and difference of the red and green signals, which cover the broadest band, are obtained. The sum R + G gives a brightness or luminance signal (Y, for yellow, somewhat misleadingly). The difference signal R-G distinguishes strongly between colors near red and green. Another chrominance signal is obtained as the difference between Y and blue (Y-B). The information processing advantage of combining and comparing color signals right at the retina is that much less noise is sent to the brain. Separate red and green signals would be nearly identical in many cases and difficult to discriminate after passing through many neurons. And the brightness range of the R-G signal is much smaller and therefore more in the capabilities of neurons to transmit. The chrominance and luminance channels have been shown to do spatial frequency analysis as well as other feature detection—both independently.

Color vision has persistence in that colors can be separately presented to the eye for short intervals, and the eye perceives the mixture. Distinguishing two colors that alternate rapidly is impossible above a certain rate of flicker (about 40 to 50 Hz). Color opponent channels do not operate faster because the delays between positive and inhibitory signals are lost. The luminance channel takes over, detecting flicker until about 70 to 80 Hz. Conversely at low flicker frequencies, color vision has better ability to detect alternation of colors than alternation of bright and dark.

Color television makes continual use of persistence. Colors are also mixed when the eye cannot resolve patches of different colors. This happens with the tiny color dots in color television as well as in everyday scenes. Color printed pictures are composed of many tiny nonoverlapping dots of a few basic colors that the eye sees combined. There does not have to be a match between the basic colors used and the sensitive ranges of the retinal pigments. Some artists (pointillists such as Georges Seurat and Manfred Schwartz) have exploited this property of color pictures and painted entire pictures with dots of pigments.

The colors of the physical visible spectrum are pure. The perceived colors are *hues*, and are measured by comparison with established standard colors.

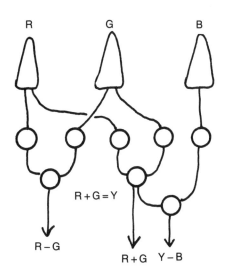

Fig. 17-2. Opponent color signal model generates two color channels (R-G, Y-B) and one luminance channel (Y = R + G) to send to the visual cortex.

About 100 hues can be distinguished. All white has been removed by the prism or grating that produces the spectrum so that there are no mixtures. *Purity* refers to a spectral color to which some proportion of white is added. In terms of intensity, a 50% pure yellow has 50% yellow and 50% white. Since there is no black light, the addition of grays to colors cannot be discussed in terms of light. Within limits, when other pure colors are mixed with a given color, the purity is reduced as if white light had been added. The color may become muddy. Without good colored lights to mix and observe, it is difficult for the reader to imagine the effects of mixing colored light. The tendency is to think in terms of mixing paints, which gives quite different results. Printers cannot directly show the results of mixing light with colored illustrations made with inks.

Saturation corresponds only approximately to purity because the sensitivity of the eye to color and to illuminance both enter into the perception of saturation. Mixtures of colors also affect saturation. The purest yellow has low saturation whereas red and blue can be highly saturated. Saturated colors appear dark but are quite different from black as increased brightness will show. Yellow is a strange color because neither yellow light nor pigment occurs in a dark saturated form. Red, green, and blue seem to lighten as saturation is reduced. The only way to equalize the brightness of yellow and other colors is to use much lower illuminance for red and blue. One way to visualize the elusive concept of saturation is to imagine diluting colored water. The basic hue used in dying the water cannot change, there is just less and less of it as water is added. (If the water affects the dye chemistry, as it does in some cases, then the hue does change.) The term *chroma* is similar to saturation in usage.

Brightness is translated from the enormous intensity range of light to paper by using a grayscale that may be based on logarithmic response (recall Fig. 16-4). When the grayscale has 9 or 10 steps, brightness may be called value. Low brightness colors, instead of being dimmer, are mixed with grays. The grays are assumed to be neutral, favoring no color. Every color can be matched to a grayscale in terms of brightness. A pure yellow would match a much lighter gray than a pure green would. It might be useful to think of detection of brightness as done with a broadband light meter that is colorblind and sees only the amount of light energy from a source or reflected from a surface.

The possible colors are infinitely many, and the eye can distinguish about 10 million of them. They can be arranged on the three dimensions (not completely independently) of hue, saturation, and brightness. People can more readily guess what colors might be mixed in a particular hue than its saturation or brightness. Most colors seem to vary with brightness, tending toward yellowish, greenish, or bluish.

When one or more of the visual pigments malfunctions or is absent, the result is a form of colorblindness. Brightness perception may be unaffected, but different colors are seen on supposedly familiar objects, and some colors may appear as gray only. The person colorblind from birth cannot understand that the colors seen are wrong. Only when, for example, red and green traffic lights appear the same, but are claimed by others to be different, is discrepancy apparent. Most cases of colorblindness are genetic, and males are most likely to be colorblind, with 8% incidence.

In the most common anomalous trichromatism, the three cone pigments seem to be present, but one or two work at much different levels, so that color mixtures are seen abnormally. A "green weak" person needs much more intense green light than others do to obtain similar visual effects. In a common form of dichromatism only yellow and blue are seen as such. Red, green, and purple are seen as grays. The very rare monochromatism is essentially rod vision. The cones are inoperative and no colors are distinguished. Acuity is poor because the cones at the fovea are not working. The colored filters used with traffic lights have been chosen to minimize confusion to the largest number of partially colorblind people.

Color vision has been verified in a wide range of animals, even fish that live in a blue-green world. Deep water fish have two-color vision (green and blue) that improves the detection of contrast. Shallow-water fish must see at a variety of brightness levels, so the high brightness adaptation of cone vision is an advantage. Near the surface fish need three-color vision, except in swamp water where only red light penetrates to the shallow bottom. As indicators of evolution, ocean animals provide evidence that two-pigment vision developed first, followed later by two-color vision, then three-color. An animal in an environment of limited color such as the deep ocean tends not to have cone cells sensitive to unavailable colors. Such cones would take up space in the retina, reducing acuity instead of providing visual information. Adaptation has also produced near-ultraviolet vision in some birds and insects, making it easier to see waxy berries. Some species of birds and fish have been shown to have four-color vision (four kinds of cone cells, including ultraviolet-sensitive cones near 370 nanometers).

17.2 COLOR MIXING AND COLOR CHARTS

Colored lights combine by addition. All the colors of the spectrum combine to form white. Most of the discussion here pertains just to visual perception—what result appears to the eye when colors are manipulated. Just red, green, and blue light, if not unusually odd hues, will combine to appear white. Almost any two carefully chosen colored lights will combine to appear white. These pairs are *complementary* colors and may be called opposites. A limited range of color can also be mixed from just two carefully selected colors, so the use of three primaries is not an immutable law. Obviously more than three colors can be used for mixing, too.

Pigments, inks, and dyes show color by absorbing other colors. If several pigments are combined, the result is not white but nearly black, usually a muddy gray. Mixing red, green, and blue paints together produces a mixture which absorbs just about all colors. Thus mixing material colors results in subtraction of colors (although some addition occurs also, to confound the results). "Mixing" can have two different meanings—mixing the colors to find the subtractive result, and mixing the actual materials. The best way to see the theoretical results of mixing is to superimpose the colored surfaces as with a rotating multi-colored wheel. The practical results

are obtained only by trial-and-error mixing. Mixing two rather dissimilar (but not complementary) colors —e.g., violet and orange as in Fig. 17-3—also results in a muddy gray-brown mixture without much definite color. Colored transparent liquids, plastics, and glasses are not exceptions to subtractive mixing. All are filters which remove all colors except the color one sees.

The principal colors used in a mixture are called *primary* colors. For light they are taken to be red (actually a red orange), green, and blue (actually a bluish purple). At equal purities, these should add to white. Within limits a mixture of colored lights, such as red and green to form yellow, cannot be distinguished from the pure spectral yellow. The requirement is that the same kinds of cone cells be used to detect the mixture or the pure color.

Pigment colors subtract to a dark gray. Theoretically the artist can mix almost any color from red, yellow, and blue, which approximate the subtractive primary colors (see center of fig. 17-3.). White and black are needed to control brightness. A good set of subtractive primary colors occurs when the additive primaries are combined in pairs: red and green (light) produce yellow (sounds unbelievable until you try it); red and blue produce magenta; blue and green produce cyan. The brightnesses and saturations must be matched, and a pure yellow will result only if the red and green are very light (like pink and mint). When pairs of the subtractive primaries are mixed, the additive primaries result: magenta and yellow produce red; yellow and cyan produce green; magenta and cyan produce blue. The two sets of primaries (really, any two sets) can be found from a chromaticity chart such as the one discussed shortly.

What about brown? Have you ever seen brown light? It is not on the spectrum, yet movies projected with light do show a full range of brownish hues. Certain mixtures of dyes in film create brown light when white is passed through. Brown as produced with pigments, dyes, and inks may come from a single material or from a mixture of colored materials. On a color chart, it can be a mixture of orange and gray. Olive browns occur when yellow is darkened. Theoretically, a highly saturated yellow would appear brownish as a pigment, and when diluted with white would gradually brighten until the familiar yellow appears. But no such pigment has been found. There are browns with reddish, orangish, yellowish, even greenish and bluish casts. Brown might be thought of as tinted gray.

Another kind of mixing is optical or visual mixing, produced on the retina by presenting two, possibly three, colors in adjacent regions. This mixing is exploited on color video screens (which have red, green, and blue luminescent dots), in color pictures printed in dots, and by some artists who paint in dots. The mixing can be by rapid superposition, as on a spinning wheel of color segments. Whether the dot images are larger than the cone cells or smaller (so that several colors impinge on one cone cell) the effect is that the additively mixed color is seen. The advantages and disadvantages of this color mixing were first expounded by Michel Chevreul in the 1820s. Applications included tapestries, where brightly colored threads placed together were seen as mixed from a distance, rendering the colors duller.

To visualize the problems of discussing color in a printed book, consider the typical three-color mix of light widely shown in illustrations. The physical primary colors of red, green, and blue are often shown overlapping to form white in a common color diagram. On the printed page, this is faking because only colored lights shining on a white screen can give such a result. On paper, red, green, and blue combine to a muddy dark gray.

The hue dimension is often closed by joining the red end through a series of red-violet mixtures to the violet end. Isaac Newton was the first to close the color circle about 1666. (Some reserve the term purple to denote red-violet mixtures.) The spectrum naturally indicates other pure mixtures of adjacent colors (blue-green, green-yellow, etc.). With yellow as a subtractive primary, mixtures of red, yellow, and blue around a triangle produce the *secondaries* orange, green, and purple, resulting in six colors around a hexagon. The color wheel is apparent when adjacent pairs of these are mixed to produce six more colors for a total of 12 (Fig. 17-3). Complemen-

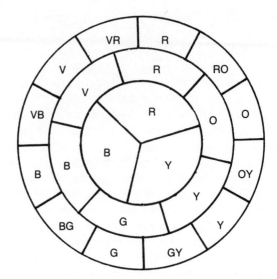

Fig. 17-3. Simple color wheel built from adjacent combinations of color pigments (subtractive mixing). A color wheel for light would use green instead as a primary. Complementary colors appear as opposites.

tary colors occupy opposite positions across the wheel.

In the *colorimeter* (Fig. 17-4), colors are matched by adjusting three colored lights (R, G, B) until the mixture appears the same to one eye as the test sample does to the same eye. The intensities of the lights are calibrated and adjustable. Color filters used to obtain colored light from white are transparent materials containing dyes. The bandpass should be as narrow as possible, so that all but a narrow range of color is absorbed. A transparent red material can have a wide or narrow bandpass, depending on its cost. For precision work, colored light is selected from a narrow part of the spectrum with a diffraction grating.

A modern and technical arrangement of colors is the *chromaticity diagram* (Fig. 17-5) as developed by the CIE (Commission Internationale de l'Eclairage). On two axes representing proportions of red and green, all possible colors obtained by mixing are shown. The basis is a three-color theory, and blue is assumed to be the remaining color needed to make up 100%. The data were obtained by measurement with colorimeters using selected

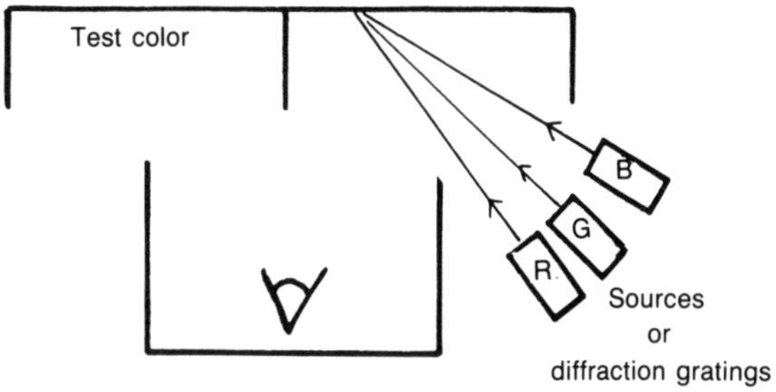

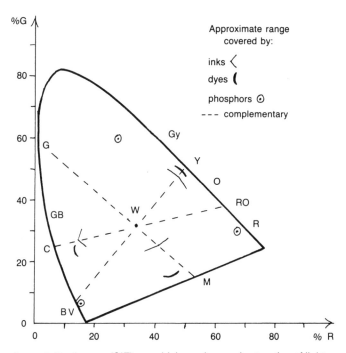

Fig. 17-4. A colorimeter is used to find what mix of primary colors duplicates an unknown or test color.

Fig. 17-5. A sketch of the chromaticity diagram (CIE), on which any hue and saturation of light or pigment is represented by a mixture of red, green, and blue. The amount of blue is determined from the combination of red and green. This diagram has no brightness scale.

standard red, green, and blue lights. The colors of the spectrum (fully saturated) appear around the outer horseshoe curve. Zero red leaves green (and blue), and zero green leaves red (and blue). The base of the horseshoe is closed with a straightline along which nonspectral mixtures of red and violet appear. At the center is white (W).

Complementary colors are found on straight lines through W. Commonly used ones complemen-

tary to red, green, and blue are cyan (C), magenta (M), and yellow (Y), respectively. Saturation increases radially outward from white. Because yellow cannot be as saturated, usable yellow appears close to the boundary. Most browns are not shown because mixtures with gray are omitted. It is impossible to reproduce the full range of colors in print (inks), with film (dyes), or in video (phosphors). These limitations are shown within the horseshoe curve. The chromaticity diagram is the closest there is to a psychophysical color theory. It has predictive power and allows limited calculation of color content. On the other hand, it shows no mixtures with gray and therefore lacks a brightness dimension. Any variations in brightness are due to the way various saturations of different hues appear.

Printers, designers, and others needing reproducible control over colors have special color charts with chips of many different colors varying in hue, saturation, and value (brightness). The charts are produced for the particular materials (medium) and show what colors result from given combinations of inks and dyes. The range is far beyond the chips widely used for house paints. The best way to measure and reproduce color is to match with chart colors in known illumination. Use of good white light is necessary for seeing color chips accurately. Color charts are expensive because each color is separately mixed and applied in printing, instead of three or four inks used to represent all colors.

The brightness, hue, and saturation depend on the surface reflectivity of sample color chips, which would best have a dull (matte or flat) finish. Color chips are arranged in various ways in three dimensions. Colors formed by adding white to saturated hues are called *tints*. Darkened colors formed by adding black are called *shades*. Colors formed by adding any shade of gray are called *tones*.

An early color chart system is the Munsell system. Hue, chroma (similar to saturation), and value (brightness on a scale of one to nine) are graphed in three dimensions as shown in Fig. 17-6. Value varies vertically from black to white, hue varies circularly around the central axis (with violet to red given the same step size as, say, violet to blue), and chroma is shown radially outward. Fully saturated colors are farthest from the axis. Unlike the spectrum, the range from red to yellow is brief while the range from green to purple occupies half the circular scale. The steps between color chips have a measured perceptual basis, being chosen to seem equal in size (in so far as equal steps can be applied to variations in color). Chroma is specified in steps from 1 (neutral) to as high as 16. Hue is broken into 100 steps or gradations (surely an arbitrary num-

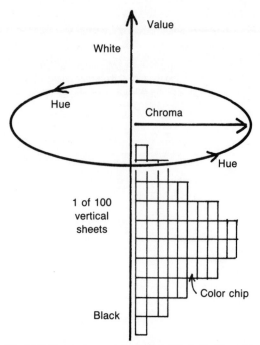

Fig. 17-6. The arrangement of color chips in the Munsell color system. Chroma is similar to saturation, and value is brightness.

ber). A sheet of chips which vary in chroma and value is made for each hue. There are more distinguishable chroma levels at moderate values, so the sheet bulges in the middle. When arranged in a circular "tree," the vertical sheets approximate a ragged sphere. Yellows are available only in rather unsaturated forms, so the tree is indented in the yellow region and bulges elsewhere. Like all printed charts, however manufactured, the Munsell must fall far short of reaching full saturation in most hues. Sheets are truncated for lack of sufficiently brilliant inks.

The Ostwald color tree has an arrangement similar to the Munsell tree but physical variables of color, luminance, and purity are used. More space is given to the red-yellow region so that complementary colors are direct opposites across the tree. Color is broken into 30 steps, so there are 30 vertical sheets. The most pure (saturated) hue is at the outer edge of the sheet, as before. Each chip has a numerical code telling how much color, white, and black are needed to make it.

17.3 SURFACES IN COLOR

The colored surfaces we see in everyday life are the result of white or near-white light reflecting diffusely from the surfaces. In the process some parts of the spectrum are absorbed, leaving the rest to reach the eye. A surface which appears green has absorbed mostly red, blue, and violet. As discussed in Chapters 9 and 12, atoms and molecules absorb light when the wavelength matches resonances of their electrons. Most atoms and molecules have electron resonances in the ultraviolet and molecular vibration resonances in the infrared and thus should appear colorless. In the visible region those atoms and molecules which absorb typically have several electrons that are easily rearranged within the atomic structure so the energy levels are closely spaced. Dyes are molecules made of chains or loops of atoms. The electrons can move rather easily along the chain and therefore have closely spaced energy levels.

The texture of the surface strongly affects its apparent color. Color is best seen when the surface is dull and has rough texture. Little white light is reflected, and the incident light is given maximum opportunity to interact with surface atoms to be absorbed. Wet surfaces, whether with water or another solvent, are more reflecting because the roughness is partly filled in. The colors appear brighter and sometimes lighter. A highlight on an object—a bright spot, perhaps of white light—is sometimes the only clue the visual system needs about the texture of the object. With a highlight the object is perceived as glossy. Without it the same object appears to have a dull finish.

Many contrast effects occur in color vision. Perceived color is strongly affected by nearby hues, particularly if there is a discernible boundary between the colors (an abrupt change in color). Regardless of hue, the brightness across boundaries is compared. The shift is toward a complementary color. A blue with a green background acquires a reddish tint (in the mind). A blue with a red background acquires a greenish tint. Conversely, colors that seem the same when far apart will show obvious differences when side by side, particularly if no border is visible and the surfaces have the same flatness.

When two colors of the same hue and similar saturation are compared, the vision system exaggerates the difference by imposing a complementary tint on the one with less saturation. (This is a clue, by the way, that saturation is a fundamental variable that can approach a maximum.) When a scene in two or more colors of the same brightness (*equiluminance*, or contrast in color only) is viewed, one discovers that most noncolor information is lost. Depth clues, detail, movement, and feature organization are difficult to discern, because the color information follows different paths of neural processing than the paths needed for these features.

When two samples of the same hue and saturation are put side by side with different colors bordering them, the two parts appear to have different saturations. For example, blue bordered by white appears less saturated than blue bordered by black. This is called the *spreading effect*. In this example, the effect is from brightness rather than hue. Contrast effects occur even when one color is gray. Making the region surrounding a bright orange a light gray depresses the shade of the orange until it appears almost brown.

Color contrast effects are probably due to inhibitory connections of cone cells to surrounding cone cells, thus requiring substantial saturation to see. Complementary effects seem built in. Afterimages are in complementary colors. A patch of light color seems even brighter at the edges when bordered by a dark color. A green patch seems to have a red edge (which can be seen only if the border is light). On the other hand, the boundary of two well-saturated complementary colors seems muted because cone cells along the image of the boundary are averaging the colors to white. Research indicates that the retina has double-opponent cells that respond differently to a red spot surrounded by green, a green spot surrounded by red, all red, all green, and combinations with black or white. The overall effect is to be very sensitive to boundaries between contrasting colors.

Random dot experiments in color show that the eye-brain system tends first to cluster colors that are neighbors in the spectrum. During analysis of a complex scene of red, yellow, green, and blue dots, the associations of red and yellow are seen together, and the blue and green are likewise associated.

In subjective interpretations of hue, the spectrum is divided into two parts, warm (for red-orange-yellow) and cool (for green-blue-violet). These are very loose terms and avoided here. They are intended to refer to complementarity. A cool color causes an adjacent color to appear warm—tinted with red even if it is also a cool color. Reddish green sounds like a contradiction until one stares at blue bordered by green and sees a reddish tint on the green! A warm color causes an adjacent color to appear cool—tinted with blue. These complementary effects are the result of fatigue of particular cone cells. When blue cells are looking at blue too much, a complementary signal of red begins. A neuron signaling almost any feature in vision soon causes a signal for the opposite feature.

There are many empirical results about color harmony—that is, what colors appear compatible together. Various systems have been developed that are sometimes called color theories without adequate justification. Hues close together and hues related as complements seem to go together. Scientific basis for these judgements is yet to be established, since so little is known of color vision at the processing stage. Artists tend to explore and exploit the limits of what looks right and what clashes. The relative areas devoted to given colors, and the relative saturations and brightnesses also affect harmony. Even the amount of area devoted to one color affects the hue seen. The possibilities to explore seem to be endless.

Everyone has enjoyed the effects of reddened sunlight on the landscape at dusk. The light source is so different from white that seeing natural colors is changed dramatically. Some colors cannot be seen in some lights. For example, shining red light on a red surface whose absorption curve resembles the spectrum of the light produces no color at all. Most of the light is absorbed and the surface appears black. Situations involving both colored and white light cause the complement to the colored light to appear in shadows.

One of the most remarkable adaptations of the eye is to perceive the same color regardless of incident light widely varying in color and illuminance. Sameness is relative, and the eye is able to compare colors while compensating for variations in illumination. The sources of light we ordinarily use (incandescent, fluorescent, sun, candle, or fire) have very different spectra. Yet we have little problem seeing basic colors. Incandescent light is very weak in blue and violet yet accurate perception of these colors is possible. The problem is only apparent when a picture is taken under incandescent light using daylight film. The picture has an orange cast when developed. Fine discrimination of colors, such as in color matching work or painting, is confounded in some lights, especially from fluorescent and other mercury-based lamps. The high-pressure sodium lamps spreading outdoors are particularly deficient in spectral content and make color discrimination a challenge.

To explain the ability to see nearly correct colors in color-biased illumination, the visual system is assumed to be comparing the colors in various parts of a scene—particularly the brightness in each of the color channels at boundaries. The brain may search for the conditions under which the ratios of brightness are consistent. Differences of hue may be distinguished from differences of brightness by comparison of the chrominance and luminance channels. The color of the illumination can be deduced if tested with the variety of colors available in most scenes. A surprising discovery is that one cannot see a particular color as nearly saturated unless other strong and contrasting colors are present.

17.4 COLOR ILLUSIONS

A color illusion that begins with black-and-white material is the spinning of a wheel with black-and-white patterns on it (first discovered by Gustav Fechner). The rapid stimulation of cone cells by the alternation of light and dark leads to saturation and recovery that differs for the different types of cone cells. Colors are momentarily seen in response. Of course the light source should be white.

Another color illusion seen on a black-and-white pattern is the McCullough effect and indicative of one way visual information is processed. One looks at a color grating—black bars separated by red or green colored bars—for several minutes, studying it. When one looks at a black-and-white grating with the same orientation, the complementary color is seen. The second viewing can be days later! Moreover, looking at a red-and-black grating tilted one way and a green-and-black grating tilted another way (Fig. 17-7A) causes the complementary color to be seen when a black-and-white grating of the appropriate tilt and spacing is viewed (Fig. 17-7B). This illusion is not an afterimage and does not require staring at one place. Research on this effect has progressed slowly, finding only that the grating has a strong influence. Its spatial frequency and orientation seem to be stored in memory for long times, along with the inhibition for seeing the color. It would seem that a chemical neurotransmitter in the synapses is associated with the stimulus and lasts for days in the retinal circuitry that was used.

Color afterimages occur when a bright, nearly saturated color is seen for many seconds then followed by a white or neutral gray screen or by simply closing the eyes. The negative afterimage of a bright saturated color is the complementary color. Generally, color illusions result from overload or overuse of particular feature detection systems. Fatigue in the retina can lead to more pervasive mental fatigue if continued.

Colored materials tend to appear the same only under the same illumination. Materials that have rather different spectral response curves may appear to have the same color under special lighting conditions. These pigments, dyes, or mixtures thereof are called *metamers* and are fortunately rare. Distinctions can be made when mixtures of light are shone on the metamers. Fluorescent lighting will show differences in very stubborn metamers. Color matching has a practical as well as a research component. If two colors are to appear identical but different materials must be used, then they should be matched for the light in which they will be used.

A two-color mixture almost simulates the full range of colors. This method uses the complementary responses of the eye. For example, photographs of a scene, taken through a green filter and a red filter, provide two black-and-white pictures with slight differences in the brightness of objects. When the photograph that originated with red light is illuminated with red light and superimposed on the

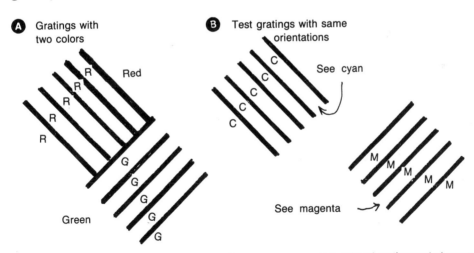

Fig. 17-7. The McCullough effect is demonstrated by staring at any one or more colored gratings set at any angles. Any time in the next few days that a black and white grating is looked at, the complementary color for the original grating of that orientation will be seen.

other illuminated with white light, a nearly full range of colors from blue through red is seen.

Studies of color vision are supported by advanced techniques such as random color dot patterns and colored gratings. As with black and white, stereopsis is a tool for locating approximately where visual processing occurs—in the brain after left and right images are brought together. Gratings are used to identify spatial frequency detection. It has been shown that the blue cones alone can detect spatial frequencies and grating orientation and send the information in a channel different from the red-green channel(s). Most studies of color vision employ red and green, and violet vision is often neglected. A more unusual experimental grating is a chromatic one. Here different colors alternate in stripes, such as red and green. Usually the brightnesses are kept the same. A grating with a smooth or sinusoidal variation in wavelength looks like a series of rainbows, or a portion thereof repeated. The brightness may be kept constant. When a constant brightness color grating is stared at, it soon becomes invisible. Lacking contrast edges, the eye is unable to recover visibility of the grating even with scanning. Chrominance channels alone seem to be insufficient for feature detection.

17.5 COLOR MATERIALS AND APPLICATIONS: DYES, INKS, PIGMENTS, AND PHOSPHORS

All materials that show color upon being illuminated are made of atoms or molecules with electron energy levels spaced such that visible light is readily absorbed. Usually the color effects are due to trace impurities rather than the bulk substance. Sometimes defects in solid materials are called color centers and have trapped electrons which can absorb and emit light in the visible range. The materials must be bulk liquids and solid—gases and finely divided crystals do not absorb color preferentially from white light. To appear as a color in ordinary circumstances, the material must have broad resonances. A sharp absorption line will have no effect on color because a limited range of wavelengths is absorbed. Every material has a spectral response curve showing how much light is reflected and ab-

sorbed at each visible wavelength. The results of mixing materials does not cause these curves to be added or combined in any simple way to give the spectral response of the mixture.

The preference of color transparencies (film, as in movies or slides) over printed paper for showing color is seen by comparing the effects of the dyes in film with the inks in paper. Dyes transmit as much as 95% of the light in the part of the spectrum where there is no intentional absorption. Inks at best reflect 85% over some of the spectrum where reflection is wanted. The relative effect at different wavelengths (the spectral response) is also greater for dyes. For example, a magenta dye transmits blue and red well and absorbs yellow and green well. A magenta ink reflects a little blue (much is wanted), reflects almost as much yellow green, and reflects much red. Transparencies also permit larger variations in brightness to be shown.

In nature, pigments and dyes seem to abound. But these colorful materials are trace molecules present for specific purposes largely unrelated to our enjoyment of a colorful scene. The reason for green plants has been discussed. Chlorophyll must absorb red and blue to carry out its natural purpose. The pigments of the biological world include the red, orange, and yellow carotenoids and the red, yellow, blue, and purple flavenoids. All animal fur and human skin is pigmented with red, tan, brown, and black melanins. Eyes are colored mainly by carotenoids and melanin. The blue comes from Rayleigh scattering from extremely small particles!

Dyes are organic chemicals usually made from coal tar. They are soluble in water or another solvent and thus break down to molecular size. They are susceptible to destruction in ultraviolet light, resulting usually in colorless compounds. Inks are made from finely ground pigment materials if intended to be permanent, or from dyes if impermanence is acceptable. Permanence of any colored material depends on its susceptibility to degradation by strong blue and ultraviolet light and by chemical reactions with air or with another pigment or dye. The selection of inks, dyes. and pigments is more chance than design. Within limits new dyes can be created with any desired colors.

Dyes are needed in color photography. As shown in Fig. 17-8, color film is made of several layers, each sensitized to a different color of light (R, G, B). More accurately, each layer contains silver halide which is sensitive to any light regardless of color (but most sensitive to blue). The silver halide is linked to a color sensitive dye. To prevent energetic blue light from affecting the sensitive green and red layers, a yellow filter layer is used above them. Still more energetic ultraviolet light has the same effect as unwanted blue light. The first color photograph was made from three photographs through red, green, and blue filters (colored solutions) by James Maxwell in 1861, although later analysis showed that his method worked for different reasons than known at the time.

During photographic developing, dyes are attached to the layers with the color for which each layer was sensitive. The dyes are deposited only where light enabled developing to produce dark silver grains. If the film is to be used as a positive transparency (slide or movie), then complementary dyes are attached instead to those parts of the film that received no light and formed no silver grains. Yellow dye goes to the blue-sensitive layer, magenta dye to the green-sensitive layer, and cyan dye to the red-sensitive layer. The dyes are not exactly complementary to the original red-green-blue sensitivities, so various corrections are made in the film, developing, and printing. The range of possible colors was indicated in the chromaticity diagram. The dyes can be so mismatched that an infrared film can be made which develops to show infrared images in false colors.

If a color negative is made, a positive color print is simply a repeat of the process, shining white light through the negative to illuminate another piece of film. Whereas color prints have a limited brightness range because white paper reflects no better than 90% and black paper absorbs no more than 90%, color transparencies can have light and dark differ by more than a factor of 100 in brightness. In either case color brightness varies in accord with the concentration of each color dye at any point.

Three general color printing methods are gravure, lithography, and serigraphy. Gravure, with the information etched into a plate, derives from intaglio (carved wood). The plate can be wrapped around a roller and run at high speed. Ink of one desired color is captured in the depressions in the surface and transferred to the paper. Three or four rollers in series and carefully aligned put three or four colors onto the continuous paper feed through. Lithography derives from the use of stone plates where the drawing is made with a grease crayon. The whole plate is wetted with water, and oil-based ink is rolled on, to adhere only on the greasy places. The modern roller plates are flat (on the small scale) and use the information in the form of ink-attracting regions. Ink and water are rolled onto the plate, then the wetted plate is rolled against a rubber transfer roller. In a second stage the transfer roller rolls the ink pattern onto paper.

The three-color approach to printing with inks or dyes (called "four-color" when black is included) usually employs magenta (midway along artificial mixtures of red and violet), cyan (a blue tending toward green), and yellow. Along with white (supplied by the paper) and black, almost any hue, saturation, and brightness can be shown on paper. Obviously yellow must be a primary color as no mixture of other colored materials gives yellow. The chromaticity diagram indicated the approximate range of colors that can be obtained with modern inks.

Color printing uses tiny dots of two, three, or more colors of ink to simulate mixtures of colors, with three (and black) the most common (Fig. 17-9). These are etched into as many printing plates. If the dots do not overlap, the color is seen by optical mixing. If they overlap, the color mixing is subtractive. The overlapping of dots requires consideration of

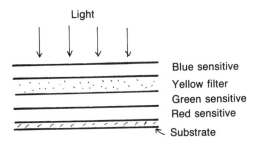

Fig. 17-8. Structure of common color film.

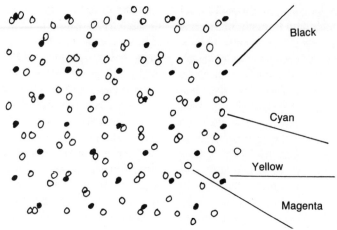

Fig. 17-9. Lithographic dots have each colored row of dots at a different angle. In this magnified diagram the spacing is exaggerated.

which order to apply inks, since semiopaque inks obscure one another. The color dots originate from photographing the subject matter three times with a camera through a screen, each time with a different color filter and a different orientation of the screen. Each negative is called a color separation, and the dots are largest where the appropriate color was most intense. The black plate is used alone in monochrome printing. Black dots are aligned at 45° to the sides of the picture as the eye is least sensitive to that orientation of grids.

The images are converted to plates by a second photographic step in which a light-sensitive resist is exposed. Where exposed it prevents the metal from being etched. The brightness range possible is less than a factor of ten. Even so, there is trouble with yellow, since yellow ink is much brighter than magenta of similar saturation. The magenta and cyan inks are usually selected light to match yellow, so black must be added for dark shades. The ink colors are fixed. Color brightness can be varied by varying the sizes of the color dots. For the most accurate color printing, the lithography is done a sheet at a time rather than continuously. It might even be done by hand on a stone as originally invented. Accurate color reproduction may require more ink colors.

Serigraphy or silk-screening is a slower, often manual process that can print colors continuously rather than in dots. Each color ink is pressed through a fine (silk) screen in places where opaque material has not been painted or pressed. There is no limit to the number of screens, each with a pattern related to one color-component of the final picture.

Pigments are not soluble and must be ground up. To make a workable paint, pigment is mixed with an almost colorless vehicle such as water, oil, acylic polymer, or another. White pigment is now commonly titanium dioxide, which is much more opaque than colored pigments. Most colored pigments are provided in nearly saturated form—very dark in appearance. A measure of their saturation is whether there is a drift in hue as white is added. Most blues, greens, and yellows change hue negligibly as they are lightened. But orange and red tend toward pink when lightened. Some red and brown pigments are based on iron oxide. Red, orange, and yellow are often based on mixtures of cadmium selenide and cadmium sulfide. Green can be obtained from chromium oxide. Iron cyanide provides a bright blue, as does cobalt oxide. Pigments were once made from minerals (best opacity) or vegetable dyes (poor opacity and permanence), or just plain dirt (clays). Now dyes are mostly artificial, more permanent but still not opaque. No deep violets were possible until coal tar dyes were developed in this range.

Artists seek very permanent dyes and pigments, which are rarely available in hues that match color charts. Available pigments are not distributed uniformly around the wheel of hues. Artists must mix pigments to obtain most colors. Charts specific to available pigment systems are available. Comparing or matching colors depends on the light used. Some pigments respond very differently to fluorescent lighting. Indirect sunlight (northlight) is the most reliable lighting for accurate color work. In some artistic methods such as watercolor, brightness or luminosity is achieved by using white paper and the paints are semitransparent. Even when painting is intended to be opaque, pigments vary widely in opacity and many layers of some are needed to hide the substrate. The use of underpaintings of similar or contrasting colors both solves and exploits this problem. The use of transparent or semitransparent color layers (glazes, not to be confused with ceramics) provides layers in which light can be reflected and filtered, so that colored illumination is built into the painting. Unfortunately, no ordinary application method produces paint films thin enough for interference effects. Techniques that exploit optical properties for artistic purposes are very numerous and are usually discovered empirically rather than by conscious study of the underlying physical laws.

Other natural and artificial colored materials are gems, glasses, and glazes. Gem stone colors come from trace metal impurities in colorless crystals—for example, 0.05% chromium ions in aluminum oxide make ruby (useful for lasers). Since emerald also has chromium, it is not the particular impurity but what energy states it has that gives the color. Clays, which are colored mainly by iron oxides, change color to reds, browns, even black (depending on the clay) when heated sufficiently to be fired. Almost any color can be glazed on by applying the appropriate mineral before firing. Copper oxide gives blue or green, iron oxide gives red, yellow, or brown, uranium oxide gives yellow, and manganese oxide gives magenta. In glass, cobalt or copper gives blue, iron gives green, selenium gives red, and nickel gives purple and brown—all as trace oxides.

A color reproduction process with growing application is color electrostatic copying. The subject matter is exposed successively to red, green, and blue light, or white light and three filters can be used. Light of a component color strikes a light sensitive plate where it causes excess charge to dissipate, leaving a region that will not attract a colored dye powder. A positive image is made in each of three complementary dyes (and black) and transferred to paper. The three colors and black are then fused to the paper with heat.

Electronic video color reproduction requires special phosphors that glow red (europium yttrium vanadate), green (zinc cadmium sulfide), and blue

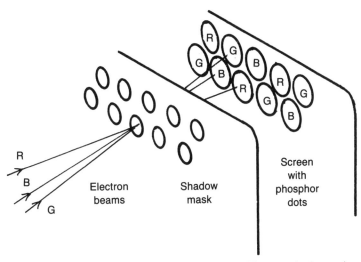

Fig. 17-10. Color video usually uses three electron beams of controlled intensities to excite three color phosphor dots at a time.

(zinc sulfide) when struck by electron beams. The phosphors are arranged in dots (or strips) over the inside of a vacuum tube (Fig. 17-10). The color video signal originates in a camera with three light sensitive scanning tubes, each with a color filter. For transmission the R, G, and B signals are combined to form a luminance signal Y. Differences R-Y and B-Y are used to generate a color signal (chrominance). Other systems have been devised. The performance of any color system can be predicted with the chromaticity diagram (recall Fig. 17-5). Any three primary colors form a triangle on the diagram, which bounds what color mixtures can be reproduced. In an RGB video system, greens and blues are given short shrift in order to cover the rest of the colors rather well. Saturated blues and greens are quite rarely seen. The diagram also indicates how three colors can be coded as two, with one color subtracted out, perhaps electronically.

Efforts continue to produce a satisfactory small, flat color display screen that needs no vacuum tube.

The light source might be a flat fluorescent plate (quasi-white) with polarizer. The screen has another polarizer and an array of tiny red, green, and blue filters. Between source and filters individually controlled liquid crystal cells retard and therefore rotate the polarization of the light either to pass or not pass through filters. The cells are electronically controlled along rows analogous to the scan of an electron beam across a video tube. There are problems obtaining bright colors and finding liquid crystals that respond sufficiently rapidly.

Color is important in communication because it provides realism in pictures and enables more efficient and complex presentation of information. The eye-brain system can process much more picture content in parallel if colored symbols are used. Ever more intricate and fascinating applications of color, lasers, and other aspects of optics can be expected as novel materials and computer-based methods are developed and the process of vision is better understood.

References

Books and articles are arranged by topic, parallel with chapter subjects where possible. If several different topics occur in one source, the book or article is listed in the general category. Books and articles are listed separately. Articles devoted primarily to simple experiments nevertheless contain considerable discussion of the subject matter. There is no practical limit to the connections between optics and other subjects. Listings either support the text, provide extra reading material near the level of the text (*denotes more advanced reading), or show selected applications that would have been discussed in the text given more space. Notes are added where needed to clarify the subject. It might seem that nearly every appropriate research book, article, and review in the world has been listed, but this is just a tiny portion of optics-related publications. Most of the journals listed here can be found in almost any library.

ScAm denotes *Scientific American*. As a resource, this bibliography lists all *Scientific American* articles pertinent to optics and visual perception back to 1960. Many of these articles have excellent photographs of optical phenomena.

ARTICLES

General (and Chs. 1, 2)

_____. ''Light,'' *ScAm* Sep 1968. (whole issue, 11 articles, all topics)

Glass, A. M. ''Optical Materials,'' *Science* 27 Feb 1987. (excellent review of new technology)

Greene, Charles. ''Glass,'' *ScAm* Jan 1961. (all about)

Javan, Ali. ''The Optical Properties of Materials,'' *ScAm* Sep 1967.

Nassau, Kurt. ''The Causes of Color,'' *ScAm* Oct 1980.

Wald, George. ''Eye and Camera,'' *ScAm* Aug 1950. (classic article)

Walker, Jearl. ''Moire effects, the kaleidoscope, and other Victorian diversions,'' *ScAm* Dec 1978. (including stereoscope)

Weisskopf, Victor. "The Three Spectroscopies," *ScAm* May 1968. (starting with atomic states, but more general)

Reflection, Refraction, and Applications, (incl. astronomical & atmospheric optics; Chs. 3, 4, 5, 6)

Carleton, N. & W. Hoffman. "The Multiple Mirror Telescope," *Physics Today* Sep 1978.

Fraser, Alistair & William Mach. "Mirages," *ScAm* Jan 1976.

Gardner, Martin. "The Curious Magic of Anamorphic Art," *ScAm* Jan 1975. (using curved mirrors)

Kachar, Bechara. "Asymmetric illumination contrast: a method of image formation for video light microscopy," *Science* 15 Feb 1985.

Kapany, Narinder. "Fiber Optics," *ScAm* Nov 1960.

Lewin, Roger. "New Horizons for Light Microscopy," *Science* 13 Dec 1985.

Maran, Stephen. "A New Generation of Giant Eyes Gets Ready to Probe the Universe," *Smithsonian* Jun 1987. (giant telescopes)

Price, William. "The Photographic Lens," *ScAm* Aug 1976. (major new methods)

Stong, C. L. "A compact short-focus telescope with spherical optical surfaces," *ScAm* Aug 1972.

_____. "Making a refractometer for the identification of liquids," *ScAm* May 1975.

Thomas, David. "Mirror Images," *ScAm* Dec 1980. (curved mirrors)

Waldrop, M. "The New Art of Telescope Making," *Science* 19 Dec 1986.

Walker, Jearl. "Anamorphic pictures: distorted views from which distortion can be removed," *ScAm* Jul 1981.

_____. "How the sun's reflection from water offers a means of calculating the slopes of waves," *ScAm* Jun 1985.

_____. "The kaleidoscope now comes equipped with flashing diodes and focusing lenses," *ScAm* Dec 1985.

_____. "Mirrors make a maze so bewildering that the explorer must rely on a map," *ScAm* Jun 1986.

_____. "The pleasures of the pinhole camera and its relative the pinspeck camera," *ScAm* Nov 1981.

_____. "Reflections from a water surface display some curious properties," *ScAm* Jan 1987.

_____. "Shadows cast on the bottom of a pool are not like other shadows." *ScAm Jul 1988.*

_____. "What is a fish's view of a fisherman and the fly he has cast on the water," *ScAm* Mar 1984. (seeing underwater)

Wave Effects—Diffraction, Interference, Interaction with Matter (Chs. 7, 8, 9)

Baumeister, Philip & Gerald Pincus. "Optical Interference Coatings," *ScAm* Dec 1970.

Bryant, Howard & Nelson, Jarmie. "The Glory," *ScAm* Jul 1974. (atmospheric display)

Chou, Tsu-wei. "Composites," *ScAm* Oct 1986. (use diffraction to measure)

*Chylek, Petr, et al. "Light Scattering by Irregular Randomly Oriented Particles," *Science* 6 Aug 1976.

Connes, Pierre. "How Light Is Analyzed," *ScAm* Sep 1968.

Darragh, P. J. et al. "Opals," *ScAm* Apr 1976. (colors from diffraction)

Greene, Charles. "Glass," *ScAm* Jan 1961.

Greenler, Robert. "Infrared Rainbow," *Science* 24 Sep 1971.

Greenler, Robert & James Mallmann. "Circumscribed Halos," *Science* 14 Apr 1972.

Greenler, Robert, et al. "Form and Origin of the Parry Arcs," *Science* 28 Jan 1977.

Greenler, Robert, et al. "The 46 Degree Halo and Its Arcs," *Science* 9 Nov 1979.

Lehn, W. "Atmospheric Refraction and Lake Monsters," *Science* 13 Jul 1979.

Lynch, David. "Atmospheric Halos," *ScAm* Apr 1978.

Nijhout, H. Frederik. "The Color Patterns of Butterflies and Moths," *ScAm* Nov 1981. (interference)

Nussenzveig, H. Moyses. "The Theory of the Rainbow," *ScAm* Apr 1977.

O'Connell, D. J. K. "The Green Flash," *ScAm* Jan 1960.

Oster, Gerald & Yasunori Nishijima. "Moiré Patterns," *ScAm* May 1963.

Pool, Robert. "Near-field Microscopes Beat the Wavelength Limit," *Science* 1 Jul 1988. (near-field pinhole detector resolves ⅙ wavelength)

Sagan, Carl. "Skies of Other Worlds," *Parade* 1 May 1988. (color illustrations and good explanation of color of skies on Venus, Mars, Jupiter, etc.)

Sawatzky, H. & W. Lehn. "The Arctic Mirage and the Early North Atlantic," *Science* 25 Jun 1976.

Shankland, R. S. "The Michelson-Morley Experiment," *ScAm* Nov 1964.

Stong, C. L. "An amateur undertakes the ultimate in mechanical precision, a ruling engine," *ScAm* Apr 1975. (to rule diffraction gratings)

_____. "How to make a ripple tank to examine wave phenomena," *ScAm* Oct 1962. (water waves to illustrate wave interference, etc)

_____. "Moiré patterns draw contours . . .," *ScAm* Oct 1973. (in 3 dimensions)

_____. "Moiré patterns provide both recreation and some analogues for solving problems," *ScAm* Nov 1964. (analog to interference)

_____. "On the making of an inexpensive diffraction grating spectrography," *ScAm* Sep 1966.

Tape, Walter. "The Topology of Mirages," *ScAm* Jun 1985.

Walker, Jearl. "A ball bearing aids in the study of light and also serves as a lens," *ScAm* Nov 1984.

_____. "The bright colors in a soap film are a lesson in wave interference," *ScAm* Sep 1978.

_____. "A homemade spectrophotometer scans the spectrum in a thirtieth of a second," *ScAm* Jan 1980. (diffraction grating type, photomultiplier)

_____. "How to create and observe a dozen rainbows in a single drop of water," *ScAm* Jul 1977.

_____. "Interference patterns made by motes on dusty mirrors," *ScAm* Aug 1981.

_____. "Mysteries of rainbows, notably their rare supernumerary arcs," *ScAm* Jun 1980.

Weisskopf, Victor. "How Light Interacts with Matter," *ScAm* Sep 1968.

Wyant, James. "Precision Optical Testing," *Science* 12 Oct 1979.

Polarization and Applications (Ch. 10)

Fergason, James. "Liquid Crystals," *ScAm* Aug 1964. (optical behavior)

*Goodby, J. W. "Optical Activity and Ferroelectricity in Liquid Crystals," *Science* 24 Jan 1986.

Heilmier, George. "Liquid Crystal Display Devices," *ScAm* Apr 1970.

Mark, Robert. "The Structural Analysis of Gothic Cathedrals," *ScAm* Nov 1972. (stress analysis with polarized light)

Stong, C. L. "How an electric field can modulate light by changing the refractivity of a crystal," *ScAm* Jul 1962. (Pockels effect)

_____. "The polariscope as a measuring instrument and as a means of creating objects of art," *ScAm* Jul 1974. (beautiful uses of polarized light)

_____. "Two methods of microscope lighting that produce color," *ScAm* Apr 1968.

Walker, Jearl. "An inexpensive homemade polarimeter can analyze optically active compounds," *ScAm* Jan 1986.

_____. "How to make dazzling photomicrographs with simple and inexpensive equipment," *ScAm* Jan 1979. (using polarized light)

_____. "More about polarizers and how to use them, particularly for studying polarized sky light," *ScAm* Jan 1978.

_____. "Studying polarized light with quarter-wave and half-wave plates of one's own making," *ScAm* Dec 1977.

_____. "What causes the color in plastic objects stressed between two polarizing filters?," *ScAm* Jun 1983.

Wehner, Rudiger. "Polarized Light Navigation by Insects," *ScAm* Jul 1976.

Quantized Light, Atoms, Molecules, and Lasers (Chs. 11, 12, 13)

Anderson, John & John Saby. "The Electric Lamp: 100 Years of Applied Physics," *Physics Today* Oct 1979. (all sorts)

Bloom, Arnold. "Optical Pumping," *ScAm* Oct 1960.

Brau, Charles. "Free Electron Lasers," *Science* 4 Mar 1988. (into visible range)

Byer, Robert. "Diode Laser-Pumped Solid-State Lasers," *Science* 12 Feb 1988.

Cross, Peter, et al. "Ultrahigh-Power Semiconductor Diode Laser Arrays," *Science* 11 Sep 1987.

Gardiner, William, Jr. "The Chemistry of Flames," *ScAm* Feb 1982. (uses laser)

Germer, Jerry. "Bright Lights," *Progressive Builder* (renamed *Custom Builder*) Dec 1986. New efficient light bulbs for residential use.

Goksu, H. et al. "Age Determination of Burned Flint by a Thermoluminescent Method," *Science* 15 Feb 1974.

Goldhaber, Alfred, & Michael Nieto. "The Mass of the Photon," *ScAm* May 1976. (no mass likely, but broad article on photon)

Hansch, Theodor, et al. "The Spectrum of Atomic Hydrogen," *ScAm* Mar 1979 (uses laser, too)

Hartmann, Sven. "Photon Echoes," *ScAm* Apr 1968.

Ingham, M. F. "The Spectrum of the Airglow," *ScAm* Jan 1972.

Johnston, T. F. Jr. "Tunable Dye Lasers," long article in *Encyclopedia of Physical Science and Technology,* Academic 1987.

Lampton, Michael. "The Microchannel Image Intensifier," *ScAm* Nov 1981.

Lempicki, Alexander & Harold Samelson. "Liquid Lasers," *ScAm* Jun 1967.

Levy, Donald. "The Spectroscopy of Supercooled Gases," *ScAm* Feb 1984. (uses laser)

Morehead, Frederick, Jr. "Light Emitting Semiconductors," *ScAm* May 1967.

Nassau, Kurt. "The Causes of Color," *ScAm* Oct 1980.

Pimentel, George. "Chemical Lasers," *ScAm* Apr 1966.

Robinson, Arthur. "A Free Electron Laser in the Visible," *Science* 3 Apr 1987.

Schawlow, Arthur. "Advances in Optical Masers," *ScAm* Jul 1963. (history of lasers)

———. "Optical Lasers," *ScAm* Jun 1961. (history of lasers)

Silfvast, William. "Metal Vapor Lasers," *ScAm* Feb 1973.

Slusher, Richart & Bernard Yurke. "Squeezed Light," *ScAm* May 1988. (good explanation)

Sobel, Alan. "Electronic Numbers," *ScAm* Jun 1973. (digital displays with light)

Thomas, David. "Electroluminescence,"

Physics Today Feb 1968.

Tsang, W. T. "The C3 Laser," *ScAm* Nov 1984 (solid state infrared)

Walker, Jearl. "A homemade mercury-vapor ion laser that emits both green and red-orange," *ScAm* Oct 1980.

———. "A homemade spectrophotometer scans the spectrum in a thirtieth of a second," *ScAm* Jan 1980.

———. "In which a Lifesaver lights up in the mouth and light takes funny bounces through a lens," *ScAm* Jul 1982. (triboluminescence, etc.)

———. "The physics and chemistry underlying the infinite charm of a candle flame," *ScAm* Apr 1978. (and spectrum)

———. "The speckle on a surface lit by laser light can be seen with other kinds of illumination," *ScAm* Feb 1982.

———. "The spectra of streetlights illuminate basic principles of quantum mechanics," *ScAm* Jan 1984.

Weinberg, Steven. "Light as a Fundamental Particle," *Physics Today* Jun 1975.

Young, Robert. "The Airglow," *ScAm* Mar 1966.

Laser Applications and Information Processing (Chs. 13, 14, 15)

Abraham, Eitan, et al. "The Optical Computer," *ScAm* Feb 1983.

Abú-Mostafa, Yaser, & Demetri Psaltis. "Optical Neural Computers," *ScAm* Mar 1987.

Alfano, R. R. & S. L. Shapiro. "Ultrafast Phenomena in Liquids and Solids," *ScAm* Jun 1973. (uses laser)

Alferness, R. C. "Optical Guided Wave Devices," *Science* 14 Nov 1986.

Anderson, Dana. "Optical Gyroscopes," *ScAm* Apr 1986.

Anderson, Rox & John Parrish. "Selective Photothermolysis: Precise Microsurgery by Selective Absorption of Pulsed Radiation," *Science* 29 Apr 1983.

Ashkin, Arthur. "Application of Laser Radiation Pressure," *Science* 5 Dec 1980.

Ashkin, A. & J. M. Dziedzic, "Optical Trapping

and Manipulation of Viruses and Bacteria,'' *Science* 20 Mar 1987. (broad report)

_____. ''The Pressure of Laser Light,'' *ScAm* Feb 1972.

Attwood, David, et al. ''Tunable Coherent X-ray,'' *Science* 14 Jun 1985. (x-ray hologram viewed with red laser light)

Baum, Gilbert & George Stroke. ''Optical Holographic Three-Dimensional Ultrasonography,'' *Science* 19 Sep 1975. (convert sound to visible)

Becker, Edwin & T. Farrar.. ''Fourier Transform Spectroscopy,'' *Science* 27 Oct 1972.

Bell, P. M. et al. ''Ultrahigh Pressure: Beyond 2 Megabars and the Ruby Fluorescence Scale,'' *Science* 2 Nov 1984. (laser measurement)

Bender, P. et al. ''The Lunar Laser Ranger Experiment,'' *Science* 19 Oct 1973.

Berger, Jon & R. Lovberg. ''Earth Strain Measurement with a Laser Interferometer,'' *Science* 16 Oct 1970.

Berns, Michael & Donald Rounds. ''Cell Surgery by Laser,'' *ScAm* Feb 1970.

Berns, Michael, et al. ''Laser Microsurgery in Cell and Developmental Biology,'' *Science* 31 Jul 1981.

Bigelow, Charles, & Donald Day. ''Digital Typography,'' *ScAm* Aug 1983. (optical storage, fourier analysis of letters)

*Bloembergen, Nicolaas. ''Nonlinear Optics and Spectroscopy,'' *Science* 4 Jun 1982. (Nobel lecture)

Boyle, W. S. ''Light-Wave Communications,'' *ScAm* Aug 1977. (first fiber optic)

Cannon, T. M. & B. R. Hunt. ''Image Processing by Computer,'' *ScAm* Oct 1981.

Christensen, Paul. ''Some Emerging Applications of Lasers,'' *Science* 8 Oct 1982.

Cook, J. S. ''Communication by Optical Fiber,'' *ScAm* Nov 1973.

Crutchfield, James, et al. ''Chaos,'' *ScAm* Dec 1986. (including visual)

Faller, James & Joseph Wampler. ''The Lunar Laser Reflector,'' *ScAm* Mar 1970.

Fassett, L. et al. ''Laser Resonance Ionization Mass Spectrometry,'' *Science* 18 Oct 1985.

Feld, M. S. & V. S. Letokhov. ''Laser Spec-troscopy,'' *ScAm* Dec 1973.

Gabor, Dennis. ''Holography 1948-1971,'' *Science* 28 Jul 1972. (Nobel lecture & great review)

Fellman, Bruce. ''They Map the Faustian Flame,'' *Smithsonian* Oct 1987. (using laser to study flames)

Gabor, Dennis. ''Holography 1948–1971,'' *Science* 28 Jul 1972. (Nobel lecture & great reviews)

Gabor, Dennis, et al. ''Holography,'' *Science* 2 Jul 1971.

Giordmaine, J. A. ''The Interaction of Light with Light,'' *ScAm* Apr 1964. (two laser beams collide in a crystal; nonlinear optics)

Giuliano, Concetto. ''Applications of Optical Phase Conjugation,'' *Physics Today* Apr 1981.

Glass, A. M. ''Materials for Optical Information Processing,'' *Science* 9 Nov 1984.

Goldstein, Charles. ''Optical Disk Technology and Information,'' *Science* 12 Feb 1982.

Hall, J. ''Stabilized Lasers and Precision Measurements,'' *Science* 13 Oct 1978.

Ham, William, et al. ''Ocular Hazard from Picosecond Pulses of Nd:YAG Laser Radiation,'' *Science* 26 Jul 1974.

Hammond, Allen. ''Optical Data Storage: Mass Memories for Future Computers?,'' *Science* 20 Apr 1973.

Harmon, L. D. & K. C. Knowlton. ''Picture Processing by Computer,'' *Science* 4 Apr 1969. (excellent)

Hemley, R. J. et al. ''Laser Techniques in High-Pressure Geophysics,'' *Science* 7 Aug 1987. (in the diamond anvil cell)

Herzenberg, Leonard, et al. ''Fluorescence-Activated Cell Sorting,'' *ScAm* Mar 1976. (uses laser)

Hiraoka, Yasushi, et al. ''Use of Charge-Coupled Device for Quantitative Optical Microscopy of Biological Structures,'' *Science* 2 Oct 1987. (for image analysis)

Hirschfeld, T. et al. ''Chemical Sensing in Process Analysis,'' *Science* 19 Oct 1984.

Holden, Constance. ''Holoart: Playing with a Budding Technology,'' *Science* 6 Apr 1979.

Holzrichter, J. et al. ''Research with High-

Power Short-Wavelength Lasers," *Science* 13 Sep 1985. (visible and ultraviolet)

Itano, Wayne, et al. "Laser Spectroscopy of Trapped Atomic Ions," *Science* 7 Aug 1987.

Jimenez, Javier, et al. "Computer Graphic Display Method for Visualizing Three-Dimensional Biological Structures," *Science* 30 May 1986.

Keyes, Robert. "What Makes a Good Computer Device?" *Science* 11 Oct 1985.

Killinger, Dennis & Norman Menyuk. "Laser Remote Sensing of the Atmosphere," *Science* 2 Jan 1987.

Kogelnik, Herwig. "High-Speed Lightwave Transmission in Optical Fibers," *Science* 31 May 1985.

LaRocca, Aldo. "Laser Applications in Manufacturing," *ScAm* Mar 1982.

Leith, Emmett & Juris Upatnieks. *Photography by Laser,"* *ScAm* Jun 1965. (holography)

Leith, Emmett. "White-Light Holograms," *ScAm* Oct 1976.

Leone, Stephen. "Laser Probing of Chemical Reaction Dynamics," *Science* 22 Feb 1985.

*Letokhov, V. "Nonlinear High Resolution Laser Spectroscopy," *Science* 24 Oct 1975.

Lewin, Roger. "New Window onto Chemists' 'Big Bang'," *Science* 11 Dec 1987 (actually the teensy bang of a reacting molecule observed with femtosecond laser pulses)

Lines, M. "The Search for Very Low Loss Fiber Optic Materials," *Science* 9 Nov 1984.

Marshall, Eliot. "Gould Advances Inventor's Claim on the Laser," *Science* 23 Apr 1982. (full history)

Maugh, Thomas. "Holographic Filing: An Industry on the Verge of Birth," *Science* 4 Aug 1978.

_____. "Remote Spectrometry with Fiber Optics," *Science* 26 Nov 1982.

_____. "X-Ray Crystallography: 3-D Structures by Optical Computing," *Science* 28 Jan 1977.

Metherell, Alexander. "Acoustical Holography," *ScAm* Oct 1969. (seen with laser)

Miller, Stewart. "Communication by Laser," *ScAm* Jan 1966.

Nelson, Donald. "The Modulation of Laser Light," *ScAm* Jun 1968.

Osgood, R. M. & T. F. Deutsch. "Laser Induced Chemistry For Microelectronics," *Science* 15 Feb 1985.

Panish, Morton & Izuo Hayashi. "A New Class of Diode Lasers," *ScAm* Jul 1971.

Patel, C. K. N. "High Power Carbon Dioxide Lasers," *ScAm* Aug 1968.

Pennington, Keith. "Advances in Holography," *ScAm* Feb 1968.

Pepper, David. "Applications of Optical Phase Conjugation," *ScAm* Jan 1986.

Phillips, William & Harold Metcalf. *Cooling and Trapping Atoms,"* *ScAm* Mar 1987.

Phillips, William, et al. "Cooling, Stopping, and Trapping Atom," *Science* 19 Feb 1988. (with laser)

Picraux, S. & L. Pope. "Tailored Surface Modification by Ion Implantation and Laser Treatment," *Science* 9 Nov 1984.

Qian, Shi-xiong, et al. "Lasing Droplets: Highlighting the Liquid-Air Interface by Laser Emission," *Science* 31 Jan 1986.

Robinson, Arthur. "Bell Labs Generates Squeezed Light," *Science* 22 Nov 1985.

_____. "Femtosecond Laser Annealing of Silicon," *Science* 19 Oct 1984.

_____. "Laser Extremes Probe Atoms and Molecules," *Science* 15 Nov 1985. (good review)

_____. "Laser Light Cools Sodium Atoms to 0.07K," *Science* 10 Dec 1982.

_____. "Multiple Quantum Wells for Optical Logic," *Science* 24 Aug 1984.

_____. "Now Four Laboratories Have Squeezed Light," *Science* 18 Jul 1986.

_____. "Sodium Atoms Trapped with Laser Light," *Science* 8 Aug 1986.

_____. "Timing Subpicosecond Electronic Processes," *Science* 28 Sep 1984.

_____. "Tunable Far IR Molecular Lasers Developed," *Science* 15 Feb 1985.

_____. "A Visible Free Electron Laser in France," *Science* 2 Sep 1983.

Ronn, Avigdor. "Laser Chemistry," *ScAm* May 1979.

Rowell, J. M. "Photonic Materials," *ScAm* Oct 1986.

Schawlow, Arthur. "Laser Spectroscopy of

Atoms and Molecules," *Science* 13 Oct 1978.

———. "Spectroscopy in a New Light," *Science* 2 Jul 1982. (Nobel lecture, lasers)

Schober, R., et al. "Laser Induced Alteration of Collagen Substructure Allows Microsurgical Tissue Welding," *Science* 13 Jun 1986. (brief broad review)

Shank, Charles. "Investigation of Ultrafast Phenomena in the Femtosecond Time Domain," *Science* 19 Sep 1986.

Shank, C. V. "Measurement of Ultrafast Phenomena in the Femtosecond Time Domain," *Science* 4 Mar 1983.

*Shannon, R. et al. "New Experimental Data on Atmospheric Propagation," *Science* 14 Dec 1979. (of light)

Sorokin, Peter. "Organic Lasers," *ScAm* Feb 1969.

Srinivasan, R. "Ablation of Polymers and Biological Tissue by Ultraviolet Lasers," *Science* 31 Oct 1986.

Stong, C. L. "How to make holograms and experiment with them or with ready-made holograms," *ScAm* Feb 1967.

———. "Infrared for the amateur: infrared diode lasers and an infrared filter," *ScAm* Mar 1973.

———. "A simple laser interferometer, an inexpensive infrared viewer, and . . .," *ScAm* Feb 1972.

———. "A tunable laser using organic dye is made at home for less than $75," *ScAm* Feb 1970.

Stroke, George, et al. "Retrieval of Good Images from Accidentally Blurred Photographs," *Science* 25 Jul 1975.

Tien, P. K. "Integrated Optics," *ScAm* Apr 1974. (laser photonics)

Vali, Victor. "Measuring Earth Strains by Laser," *ScAm* Dec 1969.

Walker, Jearl. "Caustics: mathematical curves generated by light shined through rippled plastic," *ScAm* Sep 1983. (with laser)

———. "Dazzling laser displays that shed light on light," *ScAm* Aug 1980. (and Ronchi filters)

———. "Easy ways to make holograms and view fluid flow," *ScAm* Feb 1980.

———. "More about edifying visual spectacles produced by laser," *ScAm* Jan 1981.

———. "Rainbow holograms, unlike conventional ones, can be observed in ordinary light," *ScAm* Sep 1986.

———. "Simple optical experiments in which spatial filtering removes the noise from pictures," *ScAm* Nov 1982.

———. "Wonders with the retroreflector, a mirror that removes distortion from a light beam," *ScAm* Apr 1986.

White, R. "Opportunities in Magnetic Materials," *Science* 5 Jul 1985. (magneto-optic laser recording)

Whitted, Turner. "Some Recent Advances in Computer Graphics," *Science* 12 Feb 1982. (computer calculation of pictures)

Wiersma, D. A., & K. Duppen. "Picosecond Holographic-Grating Spectroscopy" *Science* 4 Sep 1987. (with a virtual grating)

Wineland, D. J. "Trapped Ions, Laser Cooling, and Better Clocks," *Science* 26 Oct 1984.

*Yardley, James. "Tunable Coherent Optical Radiation for Instrumentation," *Science* 17 Oct 1975.

Yariv, Amnon. "Guided Wave Optics," *ScAm* Jan 1979. (uses laser)

Zare, Richard. "Laser Chemical Analysis," *Science* 19 Oct 1984.

———. "Laser Separation of Isotopes," *ScAm* Feb 1977.

Other Applications of Optics
(Science, Technology, Arts; Chs. 15, 17)

Arnon, Daniel. "The Role of Light in Photosynthesis," *ScAm* Nov 1960. (early)

Benson, D. K. et al. "Solid State Electrochromic Switchable Window Glazings," SERI Report Apr 1986.

Butler, W. L. & R. J. Downs. "Light and Plant Development," *ScAm* Dec 1960.

Czeisler, Charles, et al. "Bright Light Resets the Human Circadian Pacemaker Independent of the Timing of the Sleep-Wake Cycle," *Science* 8 Aug 1986.

Ehleringer, James & Irwin Forseth. "Solar Tracking by Plants," *Science* 5 Dec 1980.

Evans, Ralph. "Maxwell's Color Photograph," *ScAm* Nov 1961. (1861)

Frenzel, Gottfried. "The Restoration of Medieval Stained Glass," *ScAm* May 1985.

Goldberg, Leo. "Ultraviolet Astronomy," *ScAm* Jun 1969.

Govindjee & Rajni Govindjee. "The Primary Events of Photosynthesis," *ScAm* Dec 1974. (plenty of optics)

Grossweiner, Leonard. "Flash Photolysis," *ScAm* May 1960. (photochemistry)

Hamakawa, Yoshihiro. "Photovoltaic Power," *ScAm* Apr 1987.

Kevan, Peter. "Sun-Tracking Solar Furnaces in High Arctic Flowers," *Science* 29 Aug 1975.

Kingsolver, Joel. "Butterfly Engineering," *ScAm* Aug 1985.

Kolata, Gina. "Finding Biological Clocks in Fetuses," *Science* 22 Nov 1985. (set by daylight)

Kopal, Zdenek. "The Luminescence of the Moon," *ScAm* May 1965. (glow from solar particles striking surface)

Kristian, Jerome, & Morley Blouke. "Charge-Coupled Devices in Astronomy," *ScAm* Oct 1982. (electronic optics)

Levine, R. P. "The Mechanism of Photosynthesis," *ScAm* Dec 1969.

Lewy, Alfred, et al. "Antidepressant and Circadian Phase-Shifting Effects of Light," *Science* 16 Jan 1987.

Mandoli, Dina, & Winslow Briggs. "Fiber Optics in Plants," *ScAm* Aug 1984.

McCosker, John. "Flashlight Fishes," *ScAm* Mar 1977 (luminescent bacteria)

McElroy, W. D. & H. H. Seliger. "Biological Luminescence," *ScAm* Dec 1962.

Menaker, Michael. "Nonvisual Light Reception," *ScAm* Mar 1972. (biorhythm)

Miller, Kenneth. "The Photosynthetic Membrane," *ScAm* Oct 1979. (light energy converted to chemical energy in plants)

Morin, James. "Light for All Reasons: Versatility in the Behavior of the Flashlight Fish," *Science* 3 Oct 1975.

Murray, Bruce & James Westphal. "Infrared Astronomy," *ScAm* Aug 1965. (methods)

Neugebauer, G. & Robert Leighton. "The Infrared Sky," *ScAm* Aug 1968. (methods)

Newman, Eric, & Peter Hartline. "The Infrared Vision of Snakes," *ScAm* Mar 1982.

Ow, David, et al. "Transient and Stable Expression of the Firefly Luciferase Gene in Plant Cells and Transgenic Plants," *Science* 14 Nov 1986.

Rosenberg, R. et al. "Applications in energy, optics, and electronics," *Physics Today* May 1980. (including thin films)

Shkunov, Vladimir, & Boris Zel'dovich, Boris. "Optical Phase Conjugation," *ScAm* Dec 1985.

Strong, C. L. "An air flash lamp advances color schlieren photography," *ScAm* Aug 1974.

_____. "Experiments in phototaxis: the response of organisms to changes in illumination," *ScAm* Oct 1964.

_____. "Schlieren photography is used to study the flow of air around small objects," *ScAm* May 1971. (color)

Tamarkin, Lawrence, et al. "Melatonin: A Coordinating Signal for Mammalian Reproduction?," *Science* 15 Feb 1985.

Wurtman, Richard. "The Effects of Light on the Human Body," *ScAm* Jul 1975.

Youvan, Douglas & Barry Marrs. "Molecular Mechanisms of Photosynthesis," *ScAm* Jun 1987.

Visual Perception and Color (Chs. 16, 17)

Albrecht, Duane. "Visual Cortical Neurons: Are Bars or Gratings the Optimal Stimuli," *Science* 4 Jan 1980.

Alexander, Franklin. "Painting Skin the Color of Life," *Am.Art.* Mar 1987. (perhaps the greatest challenge in working with light and color)

Attneave, Fred. "Multistability in Perception," *ScAm* Dec 1971.

Bahill, A. Terry, & Lawrence Stark. "The Trajectories of Saccadic Eye Movements," *ScAm* Jan 1979.

Beck, Jacob. "The Perception of Surface Color," *ScAm* Aug 1975.

Biederman, Irving. "Perceiving Real-World Scenes," *Science* 7 Jul 1972.

Botstein, David. "The Molecular Biology of

Color Vision," *Science* 11 Apr 1986. (premature claim that color vision is fully understood)

Bower, T.G.R. "The Object in the World of the Infant," *ScAm* Oct 1971.

———. "The Visual World of Infants," *ScAm* Dec 1966.

Brindley, G. S. "Afterimages," *ScAm* Oct 1963. (to understand color vision)

Brou, Philippe, et al. "The Colors of Things," *ScAm* Sep 1986.

Brown, C. M. "Computer Vision and Natural Constraints," *Science* 22 Jun 1984.

Burgess, A. et al. "Efficiency of Human Visual Signal Discrimination," *Science* 2 Oct 1981.

Burt, Peter & Bela Julesz. "A Disparity Gradient Limit for Binocular Fusion," *Science* 9 May 1980.

Campbell, Fergus, & Lamberto Maffei. "Contrast and Spatial Frequency," *ScAm* Nov 1974. (very important)

Chen, De-Mao, et al. "The Ultraviolet Receptor of Bird Retinas," *Science* 20 Jul 1984.

Curcio, Christine, et al. "Distribution of Cones in Human and Monkey Retina," *Science* 1 May 1987.

Dobelle, W. et al. "Artificial Vision for the Blind," *Science* 1 Feb 1974.

Evans, Ralph. "Maxwell's Color Photograph," *ScAm* Nov 1961.

Ewert, Jorg-Peter. "The Neural Basis of Visually Guided Behavior," *ScAm* Mar 1974.

Fantz, Robert. "The Origin of Form Perception," *ScAm* May 1961.

Favreau, Olga & Patrick Cavanagh. "Color and Luminance: Independent Frequency Shifts," *Science* 15 May 1981.

Favreau, Olga, & Michael Corballis. "Negative Aftereffects in Visual Perception," *ScAm* Dec 1976.

Fender, Derek. "Control Mechanisms of the Eye," *ScAm* Jul 1964.

Finke, Ronald. "Mental Imagery and the Visual System," *ScAm* Mar 1986.

Fleming, Stuart. "Detecting Art Forgeries," *Physics Today* Apr 1980.

Frome, Francine, et al. "Shifts in Perception of Size After Adaption to Grating," *Science* 14 Dec 1979.

Gibson, Eleanor & Richard Walk. "The Visual Cliff," *ScAm* Apr 1960. (depth perception)

Gilchrist, Alan. "Perception of Surface Blacks and Whites," *ScAm* Mar 1979.

Gillam, Barbara. "Geometrical Illusions," *ScAm* Jan 1980.

Gogel, Walter. "The Adjacency Principle in Visual Perception," *ScAm* May 1978.

Goldberg, S. et al. "Inhibitory Influence of Unstimulated Rods in the Human Retina," *Science* 8 Jul 1983.

Goodman, Calvin. "Wilson Hurley's Color Theory and Practice," *Am.Art.* May 1986.

Gouras, P. & E. Zrenner. "Enhancement of Luminance Flicker by Color Opponent Mechanisms," *Science* 10 Aug 1979.

Green, Marc, et al. "A Comparison of Fourier Analysis and Feature Analysis in Pattern Specific Color Aftereffects," *Science* 9 Apr 1976.

Gregory, Richard. "Visual Illusions," *ScAm* Nov 1968.

Haber, Ralph. "Eidetic Images," *ScAm* Apr 1969. (photographic recall)

Harmon, Leon. "The Recognition of Faces," *ScAm* Nov 1973. (spatial frequency)

Harmon, Leon & Bela Julesz. "Masking in Visual Recognition: Effects of Two-Dimensional Filtered Noise," *Science* 15 Jun 1973. (faces)

Held, Richard & Stefanie Shattuck. "Color and Edge Sensitive Channels in the Human Visual System," *Science* 15 Oct 1971. (McCullough effect)

Hess, Eckhard. "Shadows and Depth Perception," *ScAm* Mar 1961.

Hoffman, Donald. "The Interpretation of Visual Illusions," *ScAm* Dec 1983.

Horridge, G. Adrian. "The Compound Eye of Insects," *ScAm* Jul 1977.

Hubbard, Ruth & Allen Kropf. "Molecular Isomers in Vision," *ScAm* Jun 1967.

Hubel, David. "The Visual Cortex of the Brain," *ScAm* Nov 1963. (& eye "wiring")

Hubel, David, & Torsten Wiesel. "Brain Mechanisms of Vision," *ScAm* Sep 1979.

Ingling, Carl. "Red-Green Opponent Spectral Sensitivity," *Science* 29 Sep 1978.

Johansson, Gunnar. "Visual Motion Perception," *ScAm* Jun 1975. (very good)

Jonides, John, et al. "Integrating Visual Information from Successive Fixations," *Science* 8 Jan 1982.

Julesz, Bela. "Experiments in the Visual Perception of Texture," *ScAm* Apr 1975.

_____. "Texture and Visual Perception," *ScAm* Feb 1965.

Kanizsa, Gaetano. "Subjective Contours," *ScAm* Apr 1976.

Kaufman, Lloyd & Irvin Rock. "The Moon Illusion," *ScAm* Jul 1962. (why it looks bigger near the horizon)

Kelly, D. "Disappearance of Stabilized Chromatic Gratings," *Science* 11 Dec 1981.

Kohler, Ivo. "Experiments with Goggles," *ScAm* May 1962. (eye learns to correct distortions introduced into images)

Kolata, Gina. "Color Vision Cells Found in Visual Cortex," *Science* 29 Oct 1982.

Kolers, Paul. "The Illusion of Movement," *ScAm* Oct 1964.

Koretz, Jane & Handelman, George. "How the Human Eye Focuses." *ScAm* Jul 1988.

Land, Edwin. "Experiments in Color Vision," *ScAm* May 1959.

_____. "The Retinex Theory of Color Vision," *ScAm* Dec 1977.

Land, Michael. "Animal Eyes with Mirror Optics," *ScAm* Dec 1978. (thin films)

Lappin, Joseph & Mark Fuqua. "Accurate Visual Measurement of Three-Dimensional Moving Patterns," *Science* 29 Jul 1983.

Lettvin, Jerome, et al. "What the Frog's Eye Tells the Frog's Brain," *Proc.of IRE* Nov 1959. (classic paper)

Levine, Joseph, & Edward MacNichol, Jr. "Color Vision in Fishes," *ScAm* Feb 1982.

Lewin, Roger. "Unexpected Progress in Photoreception," *Science* 1 Feb 1985.

Livingstone, Margaret. "Art Illusion and the Visual System," *ScAm* Jan 1988.

Livingstone, Margaret, & David Hubel. "Segregation of Form, Color, Movement, and Depth: Anatomy, Physiology, and Perception," *Science* 6 May 1988. (major review of latest knowledge of visual processing in brain)

MacNichol, Edward, Jr. "Three-Pigment Color Vision," *ScAm* Dec 1964.

Masland, Richard. "The Functional Architecture of the Retina," *ScAm* Dec 1986.

May, James & Halsey Matteson. "Spatial Frequency—Contingent Color Aftereffects," *Science* 9 Apr 1976.

Mays, Lawrence & David Sparks. "Saccades Are Spatially Not Retinocentrically Coded," *Science* 6 Jun 1980.

McCann, John. "Rod Cone Interactions: Different Color Sensations from Identical Stimuli," *Science* 16 Jun 1972.

Metelli, Fabio. "The Perception of Transparency," *ScAm* Apr 1974.

Meyer, Glenn & William Maguire. "Spatial Frequency and the Mediation of Short Term Visual Storage," *Science* 4 Nov 1977.

Michael, Charles. "Retinal Processing of Visual Images," *ScAm* May 1969.

Muntz, W. R. A. "Vision in Frogs," *ScAm* Mar 1964. (pioneering work)

Neisser, Ulric. "The Processes of Vision," *ScAm* Sep 1968.

Noton, David & Lawrence Stark. "Eye Movements and Visual Perception," *ScAm* Jun 1971.

O'Brien, David. "The Chemistry of Vision," *Science* 3 Dec 1982.

Oster, Gerald. "Phosphenes," *ScAm* Feb 1970. (patterns when eyes closed)

Pettigrew, John. "The Neurophysiology of Binocular Vision," *ScAm* Aug 1972.

Pinsky, L. et al. "Light Flashes Observed by Astronauts on Apollo 11 through Apollo 17," *Science* 8 Mar 1974. (in eyes, from cosmic rays)

Platt, John. "How We See Straight Lines," *ScAm* Jun 1960.

Poggio, Tomaso & Chritoff Koch. "Synapses That Compute Motion," *ScAm* May 1987.

Poggio, Tomaso. "Vision by Man and Machine," *ScAm* Apr 1984.

Pollen, Daniel, et al. "How Does the Striate Cortex Begin the Reconstruction of the Visual World," *Science* 2 Jul 1971.

Prentice, W. C. H. "Aftereffects in Perception," *ScAm* Jan 1962.

Pritchard, Roy. "Stabilized Images on the Retina,'' *ScAm* Jun 1961. (fascinating)

Probst, Thomas, et al. ''Interaction between Perceived Self-Motion and Object-Motion Impairs Vehicle Guidance,'' *Science* 3 Aug 1984. (vision & driving)

Ramachandran, Vilayanur. ''Perceiving Shape from Shading,$CA *ScAm* Aug 1988.

Ramachandran, Vilayanur, & Stuart Anstis. ''The Perception of Apparent Motion,'' *ScAm* Jun 1986.

Ratliff, Floyd. ''Contour and Contrast,'' *ScAm* Jun 1972.

Regan, David, et al. ''The Visual Perception of Motion in Depth,'' *ScAm* Jul 1979.

Regan, D. & K. Beverley. ''How Do We Avoid Confounding the Direction We Are Looking and the Direction We Are Moving?'' *Science* 8 Jan 1982.

Rock, Irvin. ''Anorthoscopic Perception,'' *ScAm* Mar 1981. (scanned images)

_____. ''The Perception of Disoriented Figures,'' *ScAm* Jan 1974. (upside down, etc.)

Ross, John. ''The Resources of Binocular Perception,'' *ScAm* Mar 1976.

Rushton, W. A. H. ''Visual Pigments and Color Blindness,'' *ScAm* Mar 1975.

_____. ''Visual Pigments in Man,'' *ScAm* Nov 1962. (earlier work)

Sagi, Dov & Bela Julesz. ''Where and What in Vision,'' *Science* 7 Jun 1985.

Sakitt, Barbara. ''Locus of Short Term Visual Storage,'' *Science* 26 Dec 1975.

Schachar, Ronald. ''The Pincushion Grid Illusion,'' *Science* 23 Apr 1976. (note critique in *Science* 2 Dec 1977)

Schnapf, Julie & Denis Baylor. ''How Photoreceptor Cells Respond to Light,'' *ScAm* Apr 1987.

Sekuler, Robert, & Eugene Levinson. ''The Perception of Moving Targets,'' *ScAm* Jan 1977. (especially good)

Shadlen, Michael & Thom Carney. ''Mechanisms of Human Motion Perception Revealed by New Cyclopean Illusion,'' *Science* 4 Apr 1986.

Shlaer, Robert. ''An Eagle's Eye: Quality of the Retinal Image,'' *Science* 26 May 1972.

Snyder, Allan. ''Optical Image Quality and the Cone Mosaic,'' *Science* 31 Jan 1986.

Stong, C. L. ''Generating visual illusions with two kinds of apparatus,'' *ScAm* Mar 1971.

Stromeyer, C. et al. ''Spatial Adaptation of Short Wavelength Pathways in Humans,'' *Science* 1 Feb 1980. (violet)

Stryer, Lubert. ''The Molecules of Visual Excitation,'' *ScAm* Jul 1987.

Teuber, Marianne. *Sources of Ambiguity in the Prints of Maurits C. Escher,'' *ScAm* Jul 1974.

Thomas, E. Llewellyn. ''Movements of the Eye,'' *ScAm* Aug 1968. (and more)

Tootell, Roger, et al. ''Deoxyglucose Analysis of Retinotopic Organization in Primate Striate Cortex,'' *Science* 26 Nov 1982. (actual map of image features—a grid—in brain)

Treisman, Anne. ''Features and Objects in Visual Processing,'' *ScAm* Nov 1986.

Vaney, David. ''Morphological Identification of Serotonin Accumulating Neurons in Living Retina,'' *Science* 25 Jul 1986. (amacrine cell patterns)

Von Derheydt, R. et al. ''Illusory Contours and Cortical Neuron Responses,'' *Science* 15 Jun 1984.

Waldrop, Mitchell. ''Computer Vision,'' *Science* 15 Jun 1984.

Walker, Jearl. ''About phosphenes: luminous patterns that appear when the eyes are closed,'' *ScAm* May 1981.

_____. ''Bidwell's ghost and other phenomena associated with the positive afterimage,'' *ScAm* Feb 1985.

_____. ''Experiments with Edwin Land's method of getting color out of black and white,'' *ScAm* Jun 1979.

_____. ''Floaters: visual artifacts that result from blood cells in front of the retina,'' *ScAm* Apr 1982. (and other illusions)

_____. ''How to stop a spinning object by humming and perceive curious blue arcs around a light,'' *ScAm* Jan 1984.

_____. ''The hyperscope and pseudoscope aid experiments on three-dimensional vision,'' *ScAm* Nov 1986.

_____. ''Illusions in the snow: more fun with random dots on the television screen,'' *ScAm* May 1980. (and Ronchi filters)

_____. ''Methods and optics of perceiving color in a black and white grating,'' *ScAm* Mar 1986.

———. "Visual illusions in random-dot patterns and television snow," *ScAm* Apr 1980.

———. "Visual illusions that can be achieved by putting a dark filter over one eye," *ScAm* Mar 1978.

———. "What explains subjective-contour illusions . . .," *ScAm* Jan 1988.

Wallach, Hans. "Perceiving a Stable Environment," *ScAm* May 1985.

———. "The Perception of Neutral Colors," *ScAm* Jan 1963.

Wallman, Josh, et al. "Local Retinal Regions Control Local Eye Growth and Myopia," *Science* 3 Jul 1987. (eye grows to achieve focus!)

Wasserman, Gerald. "Invertebrate Color Vision and the Tuned Receptor Paradigm," *Science* 20 Apr 1973.

Waterman, Talbot & Alan Pooley. "Crustacean Eye Fine Structure Seen with Scanning Electron Microscopy," *Science* 11 Jul 1980.

Wehner, Rudiger. "Polarized Light Navigation by Insects," *ScAm* Jul 1976.

Werblin, Frank. "The Control of Sensitivity in the Retina," *ScAm* Jan 1973.

White, Harvey & Paul Levatin. "Floaters in the Eye," *ScAm* Jun 1962. (diffraction causes)

Williams, David & Robert Collier. "Consequences of Spatial Sampling by a Human Photoreceptor Mosaic," *Science* 22 Jul 1983.

Wolfe, Jeremy. "Hidden Visual Processes," *ScAm* Feb 1983.

Wurtz, Robert, et al. "Brain Mechanisms of Visual Attention," *ScAm* Jun 1982.

Yellott, John, Jr. "Binocular Depth Inversion," *ScAm* Jul 1981. (relief reversal)

———. "Spectral Consequences of Photoreceptor Sampling in the Rhesus Retina," *Science* 22 Jul 1983.

Young, Richard. "Visual Cells," *ScAm* Oct 1970.

BOOKS

General (and Chs. 1, 2)

—. *Physics of Technology*, McGraw-Hill 1976. (many practical teaching modules on optical subjects)

Asimov, Isaac. *Understanding Physics: Light, Magnetism, Electricity*, Mentor 1966.

Born, Max, & Emil Wolf. *Principles of Optics*, Pergamon 1980.

Bragg, William. *Universe of Light*, Dover 1933.

*Crawford, Frank S. Jr. *Waves*, McGraw-Hill 1968.

Einstein, Albert & Leopold Infeld. *The Evolution of Physics*, Simon & Schuster 1938.

Falk, David, et al. *Seeing the Light: Optics in Nature, Photography, Vision, Holography, and Color*, Harper & Row 1986.

*Feynman, Richard, et al. *Lectures on Physics*, Addison Wesley 1964.

Gamow, George. *Biography of Physics*, Harper 1961. (history)

*Halliday, David & Robert Resnick. *Physics*, Wiley 1978.

*Hecht, Eugene & Alfred Zajac. *Optics*, Addison-Wesley 1974. (excellent)

*Jackson, John. *Classical Electrodynamics*, Wiley 1962. (very advanced)

*Jenkins, Francis & Harvey White. *Fundamentals of Optics*, McGraw-Hill 1976.

Kock, Winston. *Sound Waves and Light Waves*, Anchor 1965.

Mason, Stephen. *A History of the Sciences*, Collier 1962.

*Meyer-Arendt, Jurgen. *Introduction to Classical and Modern Optics*, Prentice-Hall 1972.

Minnaert, M. *Nature of Light and Color in the Open Air*, Dover 1954.

Rainwater, Clarence. *Light and Color*, Golden 1971. (small but no kid's book)

*Reitz, John & Frederick Milford. *Foundations of Electromagnetic Theory*, Addison Wesley 1960.

Shamos, Morris. *Great Experiments in Physics*, Holt Rinehehart Winston 1959. (some of the original optics experiments)

Waldman, Gary. *Introduction to Light: The Physics of Light, Vision, and Color*, Prentice-Hall 1983.

Weisskopf, Victor. *Knowledge and Wonder*, Anchor 1966.

Reflection, Refraction, and Applications (inc. astronomical & atmospheric optics; (Chs. 3, 4, 5, 6)

Baker, Cozy. *Through the Kaleidoscope*, Beechcliff 1985.

Boyer, Carl. *The Rainbow: From Myth to Mathematics*, Princeton UP 1987.

Edgerton, Harold & James Killian Jr. *Moments of Vision: The Stroboscopic Revolution in Photography*, MIT 1979.

Ford, Brian. *Single Lens: The Story of the Simple Microscope*, Harper 1985.

Gardner, Martin. *The Ambidextrous Universe*, Scribner 1979. (mirror reversals)

Leeman, Fred. *Hidden Images*, Abrams 1975. (anamorphic art, with experiments)

*Marchand, E. W. *Gradient Index Optics*, Academic 1978.

Welford, W. T. & R. Winston. *The Optics of Nonimaging Concentrators: Light and Solar Energy*, Academic 1978.

Wave Effects—Diffraction, Interference, Interaction with Matter (Chs.7, 8, 9)

Grafton, Carol. *Optical Designs in Motion with Moire Overlays*, Dover 1976. (for experiments)

Greenler, Robert. *Rainbows, Halos, and Glories*, Cambridge UP 1980.

Hayes, W. & R. Loudon. *Scattering of Light by Crystals*, Wiley 1978.

Heuer, Kenneth. *Rainbows, Halos, and Other Wonders*, Dodd Mead 1978. (for kids)

McCartney, Earl. *Optics of the Atmosphere*, Wiley 1976.

Wood, Elizabeth. *Crystals and Light*, Dover 1977.

Polarization (Ch 10)

Konnen, G.P. *Polarized Light in Nature*, Cambridge UP 1988/ (major reference)

Quantized Light, Atoms, Molecules, and Lasers (Chs. 11, 12, 13)

_____. *Lasers and Optical Pumping*, reprints, AIP 1965.

_____. *Quantum and Statistical Aspects of Light*, reprints, AIP.

Andrade e Silva, J. & G. Lochak. *Quanta*, World UL 1969.

*Leighton, Robert. *Principles of Modern Physics*, McGraw-Hill 1959.

Reid, R. W. *The Spectroscope*, Signet 1965. (basic)

Laser Applications and Information Processing (Chs. 13, 14, 15)

Beck, A. H. W. *Words and Waves: Introduction to Electrical Communications*, World UL 1967. (information theory)

Brill, Thomas. *Light: Its Interaction with Art and Antiquities*, Plenum 1980.

Cannon, Don & Gerald Luecke. *Understanding Communications Systems*, Texas Instruments 1980.

*Cathey, W. *Optical Information Processing and Holography*, Wiley 1974.

*Collier, Robert, et al. *Optical Holography*, Academic 1971.

*Francon, M. *Laser Speckle and Applications in Optics*, Academic 1979.

Gottlieb, Herbert. *Experiments Using a Helium-Neon Laser*, Metrologic 1973.

Greenberg, Donald. *The Computer Image*, Addison Wesley 1982. (includes color)

Heavens, O. S. *Optical Masers*, Wiley 1964.

Hecht, Jeff & Dick Teresi. *Laser: Supertool of the 1980s*, Ticknor & Fields 1982. (hundreds of laser uses)

Kallard, Thomas. *Exploring Laser Light*, Optosonic 1977.

—. *Laser Art and Optical Transforms*, Optosonic 1979.

Kaspar, Joseph & Steven Feller. *The Hologram Book*, Prentice-Hall 1985.

Kock, Winston. *Lasers and Holography*, Dover 1981.

Pierce, John. *Symbols, Signals, and Noise*, Harper & Row 1961.

Ready, J. F. *Industrial Applications of Lasers*, Academic 1978.

*Taylor, Charles. *Images: A Unified View of Diffraction and Image Formation . . .* , Crane Russak 1979.

Unterseher, Fred, et al. *Holography Handbook*, Ross 1982.

*Yu, Francis. *Optics and Information Theory*, Wiley 1976.

Visual Perception and Color (Chs. 16, 17)

_____. *Colorimetry*, reprints, AIP.

Beck, Jacob. *Surface Color Perception*, Cornell UP 1972.

Birren, Faber. *Color Perception in Art*, Schiffer 1986.

Birren, Faber, ed. *The Color Primer*, Van Nostrand Reinhold. (on Ostwald)

_____. *The Elements of Color*, Van Nostrand Reinhold 1970. (Itten)

_____. *Munsell: A Grammar of Color*, Van Nostrand Reinhold 1969.

Birren, Faber. *Principles of Color*, Van Nostrand Reinhold 1969.

Boynton, Robert. *Human Color Vision*, Holt Rinehart Winston 1979.

Coren, Stanley & Joan Girgus. *Seeing is Deceiving: The Psychology of Visual Illusions*, Erlbaum 1978.

Cornsweet, Tom. *Visual Perception*, Academic 1970.

Dars, Celestine. *Images of Deception*, Phaidon 1979. (illusions in art)

DeGrandis, Luigina. *Theory and Use of Color*, Abrams 1986.

Ernst, Bruno. *Adventures with Impossible Figures*, Tarquin 1986.

_____. *The Magic Mirror of M. C. Escher*, Random House 1976.

Fleming, Stuart. *Authenticity in Art: The Scientific Detection of Forgery* Crane Russak 1976.

Frisby, John. *Seeing: Illusion, Brain, and Mind*, Oxford 1980

Fromm, Robert. *Science, Art, and Visual Illusion*, Simon & Schuster 1971.

Fry, Roger. *Seurat*, Phaidon 1971. (pointillist painter)

Gibson, James. *The Ecological Approach to Visual Perception*, Houghton Mifflin 1979.

Gregory, R. L. *Concepts and Mechanisms of Perception*, Scribners 1974.

_____. *Eye and Brain*, World UL 1978.

Gregory, Richard. *The Intelligent Eye*, McGraw-Hill 1970. (with 3-D examples)

Gregory, R. L. & E. H.m Gombrich. eds. *Illusions In Nature and Art*, Scribners 1975.

Hubel, David. *Eye, Brain, and Vision*, Freeman 1987.

Hunter, R. *The Measurement of Appearance*, Wiley 1975.

Judd, D. B. & G. Wysecki. *Color in Business, Science, and Industry*, Wiley, 1975.

Julesz, Bela. *Foundations of Cyclopean Perception*, U.of Chicago 1971. (fantastic, with hundreds of red-green stereoptic pictures and viewer)

Kranz, S. *Science and Technology in the Arts*, Van Nostrand Reinhold.

Lanners, Edi, ed. *Illusions*, Holt Rinehart Winston 1973. (with experiments)

Lythgoe, J. *The Ecology of Vision*, Oxford UP 1979.

Marr, David. *Vision: Computational Investigation into Human Representation*, Freeman 1982.

Mauldin, John. *Perspective Design*, Van Nostrand Reinhold 1985. (mathematical, physical, and perceptual theory and practice of perspective)

Mayer, Ralph. *The Artist's Handbook of Materials and Techniques*, Viking 1970.

Nassau, Kurt. *The Physics and Chemistry of Color*, Wiley 1984.

Pellegrino, Ronald. *The Electronic Arts of Sound and Light*, Van Nostrand Reinhold 1983.

Pietsch, Paul. *Shufflebrain*, Houghton Mifflin 1981. (holographic model)

Pirenne, M. *Optics, Painting, and Photography*, Cambridge UP 1970.

_____. *Vision and the Eye*, Chapman & Hall 1967.

Rose, Albert. *Vision: Human and Electronic*, Plenum 1973.

Rossotti, Hazel. *Colour: Why the World Isn't Gray*, Princeton UP 1983. (excellent)

Samuels, Mike & Nancy. *Seeing with the Mind's Eye*, Random House 1975. (visual perception at the highest levels, the psychological side of vision)

Stone, Jonathon. *Parallel Processing in the Visual System*, Plenum 1983.

Teitelbaum, Philip. *Physiological Psychology*, Prentice Hall 1967.

Turner, Harry. *Triad Optical Illusions*, Dover 1978. (designs)

Vero, Radu. *Understanding Perspective*, Van Nostrand Reinhold 1980.

Wasserman, Gerald. *Color Vision*, Wiley 1979.

White, Lawrence & Ray Broekel. *Optical Illusions*, Watts 1986. (for kids)

Index

abn-8230

Other Bestsellers From TAB

☐ **SUPERCONDUCTIVITY—THE THRESHOLD OF A NEW TECHNOLOGY—Jonathan L. Mayo**

Superconductivity is generating an excitement in the scientific world not seen for decades! Experts are predicting advances in state-of-the-art technology that will make most existing electrical and electronic technologies obsolete! This book is one of the most complete and thorough introductions to a multifaceted phenomenon that covers the full spectrum of superconductivity and superconductive technology. 160 pp., 58 illus.

Paper $14.95   **Hard $18.95**
Book No. 3022

☐ **VIOLENT STORMS—Jon Erickson**

From the *Discovering Earth Science Series*, this book provides up-to-date information on recurring atmospheric disturbances. The internal and external mechanisms that cause weather on the Earth and the way these forces come together to produce our climate are examined. Many photographs, line drawings, and tables, as well as a complete glossary make this engrossing book informative, entertaining and easy to read. 240 pp., 190 illus.

Paper $19.95   **Hard $24.95**
Book No. 2942

☐ **FIBEROPTICS AND LASER HANDBOOK—2nd Ed.—Edward L. Safford, Jr. and John A. McCann**

Explore the dramatic impact that lasers and fiberoptics have on our daily lives—PLUS, exciting ideas for your own experiments! Now, with the help of experts Safford and McCann, you'll discover the most current concepts, practices, and applications of fiberoptics, lasers, and electromagnetic radiation technology. Included are terms and definitions, discussions of the types and operations of current systems, and amazingly simple experiments you can conduct! 240 pp., 108 illus.

Paper $19.95   **Hard $24.95**
Book No. 2981

☐ **VOLCANOES AND EARTHQUAKES—Jon Erickson**

The theory of Earth's creation through dynamic and destructive compulsion—and how those same energies continue to affect our planet—is examined in this compelling book. Discover how scientists predict catastrophes and the ways in which these violent events actually benefit Earth. 304 pp., 71 illus.

Paper $17.95   **Hard $22.95**
Book No. 2842

Send $1 for the new TAB Catalog describing over 1300 titles currently in print and receive a coupon worth $1 off on your next purchase from TAB.

*Prices subject to change without notice.

To purchase these or any other books from TAB, visit your local bookstore, return this coupon, or call toll-free 1-800-233-1128 (In PA and AK call 1-717-794-2191).

Product No.	Hard or Paper	Title	Quantity	Price

☐ Check or money order enclosed made payable to TAB BOOKS Inc.

Charge my ☐ VISA ☐ MasterCard ☐ American Express

Acct. No. _____ Exp. _____

Signature _____

Please Print

Name _____

Company _____

Address _____

City _____

State _____ Zip _____

Subtotal	
Postage/Handling ($5.00 outside U S A and Canada)	$2.50
In PA add 6% sales tax	
TOTAL	

Mail coupon to:

TAB BOOKS Inc.
Blue Ridge Summit
PA 17294-0840

BC